EMPIRE'S LABOR

EMPIRE'S LABOR

The Global Army
That Supports U.S. Wars

Adam Moore

CORNELL UNIVERSITY PRESS   ITHACA AND LONDON

This book is freely available in an open access edition thanks to TOME (Toward an Open Monograph Ecosystem)—a collaboration of the Association of American Universities, the Association of University Presses, and the Association of Research Libraries—and the generous support of Arcadia, a charitable fund of Lisbet Rausing and Peter Baldwin, and of the UCLA Library. Learn more at the TOME website, available at: openmonographs.org.

First published 2019 by Cornell University Press

Library of Congress Cataloging-in-Publication Data

Names: Moore, Adam, 1976– author.
Title: Empire's labor : the global army that supports U.S. wars / Adam Moore.
Description: Ithaca [New York] : Cornell University Press, 2019. | Includes
    bibliographical references and index.
Identifiers: LCCN 2019006498 (print) | LCCN 2019009151 (ebook) |
    ISBN 9781501716393 (pdf) | ISBN 9781501716386 (epub/mobi) |
    ISBN 9781501742170 | ISBN 9781501742170 (pbk)
Subjects: LCSH: Private military companies. | Private military
    companies—Employees. | Logistics—Contracting out—United States. |
    United States—Armed Forces—Foreign countries.
Classification: LCC UB149 (ebook) | LCC UB149 .M66 2019 (print) |
    DDC 355.3/40973—dc23
LC record available at https://lccn.loc.gov/2019006498

Empire is in the details—*Catherine Lutz*

Contents

Figures

Tables

Abbreviations

AAFES	Army and Air Force Exchange Service
AES	Anglo-European Services
AFRICOM	U.S. Africa Command
AMC	Army Materiel Command
AOR	area of responsibility
CAC	Common Access Card
CENTCOM	U.S. Central Command
CIA	Central Intelligence Agency
CSL	Cooperative Security Location
CWC	Commission on Wartime Contracting for Iraq and Afghanistan
DCMA	Defense Contract Management Agency
DFAC	dining facility
DLA	Defense Logistics Agency
DoD	Department of Defense
DoS	Department of State
FOB	forward operating base
FOIA	Freedom of Information Act
FOL	forward operating location
GAO	Government Accountability Office
GCC	Gulf Catering Company
GDP	gross domestic product
GLOC	ground line of communication
HCN	Host Country National (alternative nomenclature for LN)
HNS	Host Nation Support
ICAO	International Civil Aviation Organization
ISR	intelligence, surveillance, reconnaissance
JCC-I/A	Joint Contracting Command-Iraq/Afghanistan
KAF	Kandahar Airfield
KBR	Kellogg, Brown & Root
LN	Local National
LOA	letter of authorization
LOGCAP	Logistics Civil Augmentation Program
MEJA	Military Extraterritorial Jurisdiction Act
MOU	memorandum of understanding

MSR	main supply route
MWR	Morale, Welfare, and Recreation
NATO	North Atlantic Treaty Organization
NDN	Northern Distribution Network
OCN	Other Country National (alternative nomenclature for TCN)
OFW	Overseas Filipino Worker
OUA	Operation United Assistance
PAE	Pacific Architects and Engineers
PMC	Private Military Company
POEA	Philippine Overseas Employment Administration
PPI	Prime Projects International
PWC	Public Warehousing Corporation
PX	Post Exchange
QA/QC	quality assurance/quality control
R & R	rest and recreation
RFFS	Rescue and Fire Fighting Service
RMA	revolution in military affairs
RMK-BRJ	Raymond International, Morrison-Knudsen, Brown & Root, and J.A. Jones Construction
SCCC	Saudi Catering & Contracting Company
SOCAFRICA	U.S. Special Operations Command Africa
SOF	Special Operations Forces
SOFA	Status of Forces Agreement
TCN	Third Country National
TRANSCOM	U.S. Transportation Command
UAE	United Arab Emirates
UN	United Nations
USAID	United States Agency for International Development

EMPIRE'S LABOR

MILITARY CONTRACTING, FOREIGN WORKERS, AND WAR

Telling the story of the United States in the world from the perspective of labor . . . remaps our interpretation of empire building by demonstrating its deep connection to the migratory routes and protean life strategies of the global working class.

—Julie Greene

This book is about the U.S. military's overseas operations, both recognized wars and clandestine campaigns. Or rather, it is about the labor required to sustain such operations, and the experiences of people from around the world that do it. For the present-day U.S. military empire is profoundly dependent upon a global army of labor that comes from countries as diverse as Bosnia, the Philippines, Turkey, India, Kenya, the United Kingdom, Sierra Leone, and Fiji.

Such a state of affairs represents a profound shift in how the U.S. fights its wars, with social, economic, and political implications that extend well beyond the battlefields. Consider the following events that took place a year and a half after the invasion of Iraq. On September 1, 2004, thousands of enraged Nepalese took to the streets of Kathmandu. Their target was the small Muslim community in the country. By the end of the night the city's largest mosque, along with a number of Muslim-owned businesses and dozens of labor-recruiting agencies, had been set on fire, and the offices of Pakistani and Gulf-based airlines ransacked. Seven people died, including three individuals killed by rioters who mistakenly identified them as Muslims.[1]

The precipitant of this outburst of violence was the execution the previous day of twelve Nepalese men by the rebel group Ansar al-Sunna in Iraq. The men had left Nepal a month earlier, lured by a local recruiting agency with promises of employment at a luxury hotel in Jordan. Instead, when they reached that country their passports were confiscated and they were told that they were being sent to Iraq to work on a U.S. military base for a Jordanian-based military logistics subcontractor, Daoud & Partners. If they refused to go they would be sent back to Nepal, still owing thousands of dollars in brokerage fees to the recruiting agency. On the way to their destination the convoy was attacked and they were kidnapped. Less than two weeks later they were killed, and the execution video posted online.[2]

The twelve men were in Iraq due to a remarkable change in how the U.S. supports overseas military operations. Since 2001 it has relied on a legion of private military companies (PMCs) that employ workers from around the world. The scale of this phenomenon is extraordinary. According to a November 2008 contracting census conducted by U.S. Central Command (CENTCOM), for instance, there were more than 266,000 contractors supporting military operations in its area of responsibility (AOR), which includes the Middle East and Afghanistan. This was just short of the number of troops deployed there during the same period. The total included roughly 163,000 people working in Iraq and 68,000 in Afghanistan, with the remainder located at various bases and logistics support hubs elsewhere in the region.[3]

There are several details from this report that are worth highlighting here. First, the data represented only a partial accounting of the U.S. military's reliance on contractors to support operations in Iraq and Afghanistan at the time as it did not include thousands of private security and support staff working for the Department of State (DoS), or those employed by the United States Agency for International Development (USAID), which has also grown more dependent on contractors to carry out its reconstruction and development projects in recent years.[4] Second, 2008 was the high-water mark for military contracting in the region due to the "surge" in Iraq that began the previous year.[5] But it was in no way anomalous. As figure 1.1 indicates, the number of contractors working in CENTCOM stayed above 200,000 from the beginning of 2008—when AOR-wide censuses were first tabulated—until late 2010.[6] At the end of 2013 nearly 100,000 were still at work in the region. The contracting workforce in CENTCOM bottomed out at roughly 42,000 in summer 2015. Since then it has increasing again, to more than 50,000, as the wars in Afghanistan and the Middle East drag on.

The third point concerns the composition of the contractor workforce. According to the military's estimate, only 15 percent of contractor personnel in 2008 were U.S. citizens. Much more numerous—at 47 percent—were what it refers to as Local Nationals (LNs). LNs are citizens of the country in which the work is performed, such as Afghan truck drivers delivering goods to forward operating bases (FOBs) in Afghanistan. Occasionally military documents also refer to this class of workers as Host Country Nationals (HCNs), though this is a rather less common term. The remaining contractors—roughly 100,000 people at the time—consisted of what the military calls either Third Country Nationals (TCNs) or Other Country Nationals (OCNs), the latter an alternative nomenclature that has

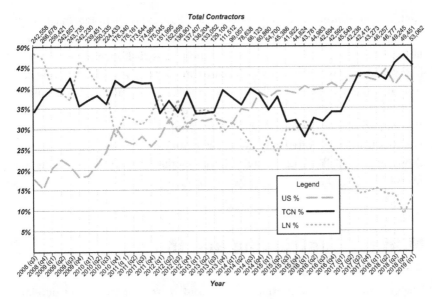

FIGURE 1.1. CENTCOM contracting statistics, 2008–2019

gained some ground in recent years. This catch-all category refers to any work-ers that are neither LNs nor U.S. citizens. The prevalence of TCN labor in 2008 was also not anomalous. As figure 1.1 shows, TCNs have represented roughly 30–45 percent of CENTCOM's contract workforce from 2008 to 2019.

The fourth important detail to consider is that just 8 percent of the military's contracting workforce was involved in providing security. This may come as a surprise to most readers because to date writing on military contracting has fo-cused on companies that provide armed security for convoys, military bases, and government personnel such as Department of State employees. Private security companies are frequently labeled mercenaries or hired guns by critics, who high-light their role in the perpetuation of human rights abuses and killings of inno-cent civilians.[7] One of the most notorious such incidents was the Nisour Square massacre in 2007, where Blackwater guards providing security for a U.S. embassy convoy shot and killed seventeen civilians in Baghdad. The voluminous academic literature on armed security PMCs tends to be state-centric and focused on pol-icy relevance. Prominent themes include the impact security contracting has upon state sovereignty and the monopoly of violence; analyses of its effectiveness; the ethical and moral implications of its use by states; and concerns about states' abil-ity to control and hold armed contractors accountable for their actions in war.[8]

Despite the focus on privatized security in the media and academia, employ-ees of armed security PMCs have constituted but a fraction of the military's contractor contingent in the wars in the Middle East and Afghanistan. In late 2008, for example, over 75 percent of CENTCOM's contractor workforce performed tasks related to logistics such as transportation, construction, maintenance, and base support. This corresponds with a 2010 military analysis of contracting data from Iraq that estimated that the ratio of contractors to uniformed personnel in the field of logistics was nearly 5:1, leading to the conclusion that "on the whole, the military is most dependent on contracted support for logistics operations."[9] Logistics workers are often employed by massive U.S. corporations like Kellogg, Brown & Root (KBR), Fluor, and DynCorp—or the multitude of subcontracting firms from around the world that they in turn rely on.

Military Contracting and the Everywhere of War

The growth of military contracting in recent years is an important development because it represents a fundamental change in how the U.S. fights it wars. What is new is not the reliance on private companies and labor to support military campaigns, which has a long history in both the U.S. military and among other armed forces, but rather the scale and scope of the phenomenon. In World War II the ratio of contractors to uniformed personnel was roughly 1:7. In Vietnam it was 1:6. In contrast, in the three largest overseas contingency operations in the past two decades—the peacekeeping missions in the Balkans (Bosnia and Kosovo) and the wars in Iraq and Afghanistan—the number of contractors has been roughly equal to or greater than the number of uniformed personnel in the theaters of operation.[10] And in Africa, where the military's presence has grown rapidly over the past decade, contractors play a central role in supporting an expanding network of drone bases, logistics nodes and clandestine Special Operations Forces (SOF). Put simply, the U.S. is now dependent on contracted labor, especially in the realm of logistics, to fight its wars.[11]

I would argue, in fact, that the U.S. military's increasing reliance on private companies and foreign labor to provide logistics support for operations around the world is as significant as the various technological innovations toward network-centric warfare over the past two decades that have been dubbed a "revolution in military affairs," or RMA.[12] Especially since, as one military analysis from 2001 notes, "RMA is predicated on a revolution in military logistics" that centers on the increasing use of private contractors.[13] Or as former army chief of staff Eric Shinseki put it in 2002, "Without a transformation in logistics there will be no transformation in the Army."[14]

The transformation of military logistics through contracting is not just operationally linked to RMA. Both have also profoundly impacted the spatiality of war, though in different ways.[15] A key claim made by political geographers and other social scientists is that RMA, combined with the U.S. response to 9/11, has led to a blurring of the traditional geographies of warfare: from defined battlefields to multidimensional and fluid urban "battlespaces"; from officially recognized combat zones to shadowy campaigns against nonstate actors in "borderlands," "ungoverned spaces," and undisclosed locations; and the development of novel forms of "lawfare" that radically redefine legal jurisdictions, detention policies, and the different classes of people that are considered "lawful targets."[16] In the evocative words of Derek Gregory, we are living in the age of "the everywhere war."[17]

Military contracting is also reshaping the geography of war by generating new political and economic entanglements, the effects of which often extend well beyond the immediate spaces of violence. These entanglements profoundly impact livelihoods, politics, and social relations in numerous communities and states around the world that are not directly involved in the various U.S. wars and military operations. Nepal's deadly violence in 2004 dramatically illustrates these distance-spanning entanglements. Put another way, the expansion of military contracting is producing what may be called the "everywhere of war."[18]

The following examples illustrate this claim. According to the Department of Labor, more than 3,380 civilians working for the U.S. military or various PMCs supporting the wars in Iraq and Afghanistan died between September 2001 and June 2018. This compares with roughly 6,950 U.S. military casualties in those wars.[19] While contractor deaths and injuries—especially foreign ones—barely register in the U.S., the same is not true of the countries that they are from. Following the deaths of the twelve workers from Nepal, the Nepalese government declared a national day of mourning. In the Philippines incidents involving workers, such as the deaths of ten men whose helicopter crashed in Afghanistan in 2009, are regularly given prominent coverage by national TV networks and newspapers.[20] Perhaps it is not surprising, then, that the plight of workers in Iraq and Afghanistan has also impacted domestic and international politics around the world. In 2004, for instance, insurgents began targeting truck convoys carrying food, fuel, and materials from Kuwait and Turkey to U.S. bases in Iraq. As deadly attacks and hostage taking of drivers mounted, India and the Philippines declared travel bans to Iraq for their citizens. They were joined by Nepal immediately following the execution of its trafficked citizens.

In each case the countries' decision to impose a ban on travel to Iraq for work was driven by domestic political considerations. Nepalese diplomats stated that the government felt "very vulnerable" following the anti-Muslim riots in the

country, owing to fears about both further domestic unrest and potential repri-
sals against the hundreds of thousands of Nepalese working in Muslim countries
in the Middle East.[21] For the Philippines the tipping point was the kidnapping of
a truck driver, Angelo de la Cruz, in July. His hostage takers threatened to kill him
if the Philippines did not remove its small contingent of troops from the country. Initially defiant, President Gloria Macapagal Arroyo, who had just won a controversial election dogged by allegations of vote rigging, eventually acquiesced to
this demand following massive protests across the country. Shortly afterward de
la Cruz was released, followed by the imposition of a travel ban.

The travel bans immediately set off alarm bells within the U.S. military due to
its dependence on workers from these labor-exporting states. It also prompted a
flurry of urgent behind the scenes diplomacy by the DoS. To give a sense of just
how dependent on TCN labor for logistical support the military was at the time,
one DoS analysis written shortly after the India and Philippines bans were announced stated:

> Coalition forces are heavily dependent on Filipino and Indian drivers and
> other logistical support personnel for the humanitarian fuels, military
> food supply and mission critical programs in Iraq. Contractors and U.S.
> military report that a fully enforced ban would cripple these operations.
> There are no readily implemented short-term workarounds to ameliorate
> the effect of a travel ban. . . . For example, Public Warehouse Company
> (PWC), the prime vendor for the supply of water and food to U.S. forces
> in Iraq, confirmed on 3 August that fully 48 percent of the firm's 1,500
> drivers are Indian and that at least 10 percent more are Filipino.[22]

Three days later the U.S. embassy in Kuwait reported that over 1,000 trucks were
stuck at the Kuwait-Iraq border, through which roughly 75 percent of goods entered Iraq at the time. It also noted that the military estimated that less than a
week's supply of food and water for troops remained in the country.[23]

Initially the U.S. tried to convince India, the Philippines, and Nepal to reverse
their travel bans, or at least exempt from them citizens that worked for military
contractors. It also promised to improve security measures for convoys, including an increase in military escort vehicles. When this approach gained little
traction—and facing an "ever-dwindling" supply of workers as other countries
imposed and pondered travel bans in the fall—it changed course and pressed Kuwait not to enforce the bans at its border crossings, which would "allow dis-
tressed contractors to move towards more normal work schedules and alleviate
the mounting logistical problems created by the travel bans."[24] At first the Kuwaiti government was resistant to this plan, especially without diplomatic cover
from countries that had imposed the travel bans, but it eventually agreed follow-

ing continued pressure from the U.S. government and Kuwaiti trucking firms that held the majority of military transport contracts for Iraq.

For the Philippines the decision to also withdraw its small contingent of troops from Iraq was even more geopolitically fraught than the imposition of a travel ban. Presenting itself as a close ally of the U.S. following 9/11, it had sent troops to Iraq as a member of the "coalition of the willing," a decision motivated in part by the lure of potential contracts and jobs that it envisioned would accompany postwar reconstruction. The decision to pull out its military contingent to secure the release of de la Cruz met with angry condemnation from other members of the coalition. Australia's foreign minister called the decision "marshmellowlike" and an "extreme disappointment," while U.S. secretary of defense Donald Rumsfeld stated that "weakness is provocative."[25] In response the U.S. withdrew its ambassador to the Philippines for consultations. It also, according to interviews with Filipino officials and workers in Iraq at the time, imposed retaliatory measures including restrictions on diplomatic personnel visiting the Green Zone and reductions in the privileges of Filipino workers on certain U.S. military bases, such as restrictions on mobility and the use of recreational facilities.

In response to criticisms from other coalition members Philippine Senate majority leader Francis Pangilinan wrote an open letter that highlighted the distinctive geopolitical situation his country faced due to its position as a major exporter of labor to the Middle East. He noted that over a million Filipino citizens were working in the region, any of which he claimed might become "targets of retaliation" if the Philippines did not withdraw from the coalition. He also observed that if other coalition partners had such a large civilian presence in the region their views about continued participation would be "starkly different."[26] As Pangilinan's comments illustrate, the position of labor-exporting states was shaped by domestic political protests in the aftermath of kidnappings and attacks on their workers in Iraq, and the fear that being seen as too closely linked to the U.S. occupation could potentially put hundreds of thousands of their citizens working elsewhere in the Middle East at risk. Therefore these states decided to distance themselves by imposing travel bans. These decisions, and the desperate attempt by the U.S. to circumvent the bans by inducing Kuwait not to enforce them at its border with Iraq, also illustrate the degree of dependency the military has on foreign labor, and the need for support—or at least indifference—from labor-exporting states in acquiring it.

The global entanglements of military contracting are also manifest in more mundane ways. Over the past two decades, for example, the economic fortunes of a number of communities in Bosnia, Kosovo, and Macedonia have been intimately linked to the growth of this phenomenon, first through employment related to peacekeeping missions in the region, later as thousands of men and women from those countries were recruited to work in Iraq and Afghanistan, and more recently when a contingent traveled to West Africa to provide support for Operation United Assistance, the 2014–15 military mission to fight the Ebola outbreak in Liberia, Guinea, and Sierra Leone.

In fact, it was while conducting PhD research on postwar peacebuilding in Bosnia that I first became aware of the impact that military contracting has had in the region and beyond. The initial encounter took place in 2005 at the University of Sarajevo's computer center, which was then located near the city's main bus terminal. One day, while checking email, I heard six students huddled around a computer talking excitedly about jobs on military bases in Iraq. Peeking over I noticed that they were looking at KBR's recruiting page. A few days later I asked two friends from the town of Brčko in northeast Bosnia about this. They had both served as interpreters for U.S. peacekeeping forces in the 1990s and said that they knew of several former interpreters who had been recruited by KBR to work in Iraq. At the time I just filed this away as a curious detail.

During further research in summer 2011, my attention was again drawn to the import of military contracting in Bosnia when several friends in Brčko discussed preparing résumés to send to recruiters in the nearby town of Tuzla, who were actively looking for workers to support Fluor's and DynCorp's expanding operations in Afghanistan. That summer residents of Brčko also mourned the death of Nenad Antić, a contractor who was killed in Afghanistan. Interest piqued, the following year I arranged to talk with a handful of individuals in Tuzla who previously worked for KBR in Iraq. My thinking at the time was to write a short article about Bosnians working on military bases in the Middle East. However, as I talked with people and delved deeper into the topic I began to realize that the significance of these dynamics extend well beyond Bosnia.

The Labor of Empire

Since the early 2000s the status of the U.S. as a modern-day empire has gone from a highly contested claim to commonplace observation among both critics and proponents. As Robert Kaplan proclaimed in 2003, "It is a cliché these days to observe that the United States now possesses a global empire. . . . It is time to move beyond a statement of the obvious."[27] To be certain, this empire looks different from earlier European examples with their vast colonial holdings. Instead of colonies, a global network of military bases provides evidence of imperial might, with one recent analysis concluding, "Although few U.S. citizens realize it, we probably have more bases in other people's lands than any other people, nation, or empire in world history."[28] Some might argue that an absence of colonies disqualifies the U.S. as an empire. But reliance on a vast network of bases in client states and allies rather than territorial colonies simply constitutes a different modality of imperial power, one based on informal rather than formal measures to exert influence on other countries.[29]

While this worldwide network of bases is considered one of the most prominent examples of U.S. imperial ambitions and military might, less attention has been paid to the global army of labor that supports overseas operations at these bases, or the political and economic entanglements that this entails. Logistics labor in particular has been overlooked in the literature dedicated to its various wars. One reason for this lack of attention is that this work seems mundane and less "mercenary."[30] Filipinos driving trucks, Kosovar Albanians cleaning latrines, and Indians cooking pancakes are not what we picture when we think of the PMC industry. Yet it is precisely these kinds of workers, these types of labor, that animate the U.S. military's overseas interventions. Consequently, it is worth asking what the world looks like when we "gaze through the looking glass at the working people and labor systems" that make U.S. empire work.[31]

Adopting a more historical perspective makes apparent that for all its unique qualities, present-day military contracting echoes earlier U.S. labor dynamics. As Julie Greene observes, "the U.S. imperial project" has "always and everywhere involved the recruitment, managing and disciplining of labor."[32] In recent years a vibrant body of research premised on the argument that "empire has a labor history" that is just beginning to be written has explored the labor that facilitated expansionary political projects following the Civil War, especially the early years of the twentieth, or "American," century, from a global workforce mobilized to build the Panama Canal; to Filipino and Puerto Rican field hands brought in to work the sugar plantations in the territory of Hawaii; to Cuban laborers who constructed and maintained the military base at Guantanamo.[33] Like today's global army of military labor, these earlier examples depended heavily on the recruitment and exploitation of foreign, nonwhite workforces. The persistence of these dynamics exemplifies what Ann Laura Stoler refers to as "imperial durabilities."[34] Thus while the following pages provide an analysis of military logistics labor in the U.S. imperial present, it is necessary to recognize that this present is also inextricably connected to its past.

Aims

This book is the product of a multiyear descent down the rabbit hole of military contracting and logistics labor. It has multiple aims. One is to outline the history of logistics outsourcing by the U.S. military, including the rapid upshift in the practice over the past two decades. Doing this necessitates situating this phenomenon within the wider context of government privatization trends in the fields of defense and intelligence, as well as the downsizing and transformation of the military following the Cold War. It also involves outlining how present-day contracting compares with logistics support supplied by camp followers, sutlers, and corporations in earlier eras.

A second goal is to illuminate the immense work involved in sustaining the U.S. overseas military empire. Over the past two decades U.S. forces have been continuously deployed fighting wars, hunting terrorists, and conducting peacekeeping and humanitarian missions across the globe. These operations, especially the wars in the Middle East and Afghanistan, are logistically intensive. Conducting them has involved the movement of a tremendous amount of goods and people along lengthy and complex supply chains, the construction and maintenance of hundreds of bases—many the size of small cities—in remote and challenging environments, and the provision of a panoply of life support services like food, laundry, showers, and billeting for uniformed personnel. All of this logistical support depends upon an army of labor drawn from around the world.

Third, this book endeavors to trace the routes and labor supply chains traversed by the military's global workforce, as well as the specific histories and present-day politics that shape them. Taken together, such pathways "represent a kind of imperial geography, tracing boundaries of an empire of mobility."[35] Given the number of countries that serve as sources of labor, these routes are varied. In some cases, as with many workers from the Balkans, one's journey began as a local hire, or LN, before following employers to military operations in the Middle East or Africa. In other instances company websites or online forums have served as an introduction to the world of military contracting. For most workers from countries like India, Nepal, and the Philippines, the recruiting process—from the role of local agents, to fees and terms of contracts, to experiences of labor trafficking—has shared characteristics with the broader recruiting assemblage that facilitates a massive labor import-export regime between wealthy Gulf petro-states and poor, Asian, labor-exporting countries. This should not be a surprise as the largest military subcontractors in Iraq and Afghanistan tend to be firms from the Middle East. But the result is that the military has in effect "imported" a host of exploitative labor practices that parallel conditions experienced by labor migrants elsewhere in the region, while at the same time deliberately exercising minimal oversight responsibility.

The fourth, and primary, goal of this book is to give voice to the agency, aspirations, and experiences of those who labor for the military—focusing specifically on foreign logistics workers whose experiences have been occluded by the overweening focus on private military security contractors. What, for example, is life and work like on a military base in a warzone? News reporting and documentaries to date—mainly by a handful of dogged journalists such as Pratap Chatterjee, Anjali Kamat, T. Christian Miller, David Phinney, Cam Simpson, Sarah Stillman, and Lee Wang—have produced a portrait of exploited laborers from South and Southeast Asia employed by subcontracting firms. This book is indebted to their work. However, while difficult and exploitative working conditions have certainly been the experience of many, life on military bases in the Middle East, Afghanistan, and Africa is rather more complex than existing accounts suggest. My research indicates that workers' experiences vary considerably and are shaped by a range of factors including nationality, gender (a not insignificant portion of workers are women), language, type of base or camp, the work one does, and what company one works for. Moreover, there is a hidden history of labor activism and worker agency on bases that has not been adequately examined to date. In addition to base life, I also examine the social, political, and economic impacts that this work has on families, and on the communities and countries that laborers come from.

Sites and Sources

The geography and scale of U.S. military contracting over the past two decades is vast. Workers from dozens of countries around the world have labored in a panoply of states across Central Asia, the Middle East, Europe, Latin America, and Africa. Obviously it is not possible to capture the full extent and diversity of this phenomenon. Any account will be partial and incomplete.

This book focuses on laborers from two countries: Bosnia and the Philippines. There are several reasons behind this choice, three of which are worth noting here. First, both countries have been significant sources of military labor over the past two decades—and even longer in the case of the Philippines, with the U.S. military continuously utilizing Filipino labor from 1898 to the present. Second, Bosnia and the Philippines are also useful for revealing the complexity and diversity of workers' experiences on military bases, while also identifying commonalities. For example, whereas most Filipinos have worked for subcontracting companies in Iraq and Afghanistan, Bosnians have by and large been employed by prime contractors like KBR, Fluor, and DynCorp. This is significant because the distinction between employment with a prime contractor or subcontractor is the most important determinant of one's pay and privileges on a base, with vast disparities between the two categories. Moreover, a focus on Bosnian and Filipino workers also offers insight into the ways in which race and nationality shape work and life on bases. Third, these countries' specific histories, including the

Philippines' decades-old labor-export economic development strategy, the colonial and client-state relationship between the U.S. and the Philippines throughout the 1900s, and the U.S. military's peacekeeping missions in the Balkans, are useful for tracing the history and development of military contracting in relation to U.S. empire and geopolitics in the twentieth century, and broader currents of transnational labor migration in recent decades.

Between 2012 and 2016 I conducted in-depth interviews with more than eighty current and former workers from Bosnia and the Philippines, interviews that in many cases included family members and multiple sessions—sessions that extended across multiple years in the case of several Bosnian workers. I also interviewed a number of recruiters and government officials in these countries. Pseudonyms are used to protect the identity of interviewees throughout the book. This is not just a matter of research ethics. Nearly every PMC employee I talked with signed a confidentiality agreement as a stipulation of employment. Therefore revealing people's identities could not just have negative effects on their future employment opportunities, but also potentially open them up to legal repercussions. This is an unlikely but not completely hypothetical risk. As I discuss in chapter 8, recruiting agencies and military subcontractors have initiated legal cases in the Philippines against former workers in Iraq who jumped to new companies prior to the completion of their original contracts, a violation of the terms of the contracts they signed.

In addition to interviews this project draws on a range of textual sources. Several Filipino workers generously shared copies of employment contracts with me, and friends in Bosnia introduced me to online forums from the region that have served as key sources of information on employment opportunities, the hiring process, and working conditions with different companies. I have also extensively mined a variety of U.S. government documents, including Government Accountability Office (GAO) reports, DoS cables, congressional testimony, and numerous military reports, websites, and investigations. A number of these documents have been obtained through Freedom of Information Act (FOIA) requests.

Organization

Thematically I divide this book into three parts. The first concerns *histories*. This section begins with a chapter that outlines the scale and scope of privatized military work in the present day, compares this with earlier practices of contracting by the U.S. military, and explains the rise of large-scale logistics outsourcing since the end of the Cold War. Following this I provide an overview of colonial and client state relations between the U.S. and the Philippines in the twentieth century, as well as the related history of reliance on Filipino labor by the U.S. military that continues to shape the recruitment of Filipinos for military work. Chapter 3 also describes the emergence of labor export as a development strategy by the Philippines starting in the 1970s, the concurrent development of labor flows between Gulf states and South and Southeast Asian countries, and links between these two processes and recruiting pathways, logistics subcontractors, and Filipino employment on U.S. military bases in Iraq and Afghanistan. Chapter 4 then examines the duality of prosperity and precarity experienced by Bosnians who have worked for the U.S. military and various contractors over the past two-plus decades. It begins by describing the economic and social significance of participation in the country's postwar "peacekeeping economy" in the 1990s, with companies providing logistics support for peacekeeping forces, or employment with one of the myriad international organizations involved in peacebuilding projects during this period. I then detail the shift to employment in warzones in the Middle East and Afghanistan as relatively privileged direct hires with Logistics Civil Augmentation Program (LOGCAP) prime contractors (i.e., KBR, Fluor, and DynCorp), followed by a discussion of the experience of both prosperity and precarity by those who have done this work.

The theme of the second part is *routes*. This includes networks, infrastructures, and practices that span and constitute the spaces through which people, information, and goods circulate. I begin in chapter 5 by describing logistics spaces and labor involved in supporting overseas operations in Iraq, Afghanistan, and, increasingly, Africa. Chapter 6 then contrasts the legal hiring processes and key nodes for recruitment and travel to and from worksites in Afghanistan and the Middle East for Bosnians, who tend to be directly hired by prime contractors, and Filipinos, who have primarily worked for subcontractors. One way I approach this is by tracing the pathways—social networks, recruiting agencies, internet forums, and company-specific application processes—that constitute these respective labor supply chains. Drawing on interviews with workers and labor brokers in the Philippines, chapter 7 examines trafficking of South and Southeast Asian workers, and the "backdoor" or underground recruitment of Filipino labor following the introduction of travel bans to Iraq and Afghanistan. This chapter also discusses the continuing problem of labor abuses—especially trafficking—and legal rationales deployed by the U.S. military to disentangle itself as much as possible from oversight responsibility.

The third part of the book focuses on *base life*. I approach this topic in a variety of ways, beginning in chapter 8 with an analysis of the hidden dynamics of labor activism on military bases in Iraq and Afghanistan, focusing in particular on three strategies: protests, strikes, and "jumping" from one company to another.

I describe the motivations of workers who engage in these actions, as well as the risks, and the coercive measures employed by companies—especially subcontractors—to suppress them. Following this, chapter 9 examines stark differences in pay, perks, and working conditions between those employed by prime contractors or subcontractors, and ways that race, nationality, and gender shape relations and hierarchies among workers and between workers and service members on bases. Chapter 10 explores the themes of family, community, and returning home. This encompasses the impact of working on bases in warzones on family life, including divorce and marriage, the economic and social impacts on communities workers hail from—which is significant given spatial concentrations of recruitment in both Bosnia and the Philippines—and difficulties in adjusting to life at home following the end of employment.

In the conclusion I step back and ask the following question: How has military contracting and the increasing reliance on foreign labor detailed in this book impacted the "American way of war"? The answer, I suggest, is that the growth of contracting has—in conjunction with technological innovations—transformed both the spatial and temporal registers of war. I have briefly discussed the changing spatial dimensions (the "everywhere war" and the "everywhere of war") above. Temporally, it has enabled what Dexter Filkins aptly refers to as the "forever war"—a ceaseless parade of military operations around the world over the past two decades in response to 9/11.[36] Crucially, military contracting transfers risk and casualties onto foreign bodies, thereby dampening domestic opposition to the pursuit of boundless war elsewhere in the world. Put another way, this global army of labor is an inextricable facet of the present-day U.S. military empire.

Part 1
HISTORIES

FROM CAMP FOLLOWERS TO A GLOBAL ARMY OF LABOR

Although there is historic precedent for contracted support to our military forces, I am concerned about the risks introduced by our current level of dependency.

—Robert Gates, former U.S. Secretary of Defense

In 2014 the world watched in horror as an outbreak of the deadly Ebola virus spread through West Africa. By the beginning of August the World Health Organization had recorded over 1,700 cases, resulting in more than 900 deaths in Guinea, Liberia, and Sierra Leone. In response it declared the outbreak a "public health emergency of international concern."[1] As the death toll rose, Ebola was increasingly framed as not just a public health crisis, but also a global security threat, culminating in a September 18 United Nations (UN) Security Council resolution declaring the epidemic "a threat to international peace and security."[2] Two days before this resolution President Barak Obama held a press conference announcing that he was deploying the U.S. military to Liberia as part of a multinational effort to stem the outbreak.[3]

The military, in turn, looked to contractors to provide critical logistics support for what became known as Operation United Assistance (OUA). The first company they reached out to was Fluor, which had just won a LOGCAP contract the previous month to provide support services in Africa.[4] Within days of Obama's announcement an advance team of Fluor employees was in Liberia conducting initial assessments.[5] Three weeks later recruiters for Fluor arrived in Tuzla, in northeast Bosnia. They quickly set up shop in the city's main hotel and began interviewing applicants. Those that passed the interviews received background checks and health exams at a local clinic and then were flown to Dubai. There they waited for necessary paperwork and watched training presentations that covered safety procedures, dangers inherent to working in warzones, life on military bases, and discussions of potential environmental and health risks. Less than two months after Obama's announcement the initial batch of recruits from Bosnia arrived in Liberia and in Dakar, Senegal, which served as the main West African logistics hub for operations. For the next five months they worked hand in hand with the military, building Ebola Treatment Units and providing an array of support services. At the peak of operations in late 2014 over 1,000 Fluor employees were supporting the OUA mission.[6]

Fluor was not the only company contracted by the military, though it was the largest. By the end of OUA over 400 contracts totaling more than $120 million had been signed.[7] Many of these were with local companies, such as a Liberian transportation firm that provided 300 trucks to move supplies across the country. All of this was by design as the "plan from the outset," according to one retrospective assessment of the logistics component of the mission, "was to attempt to contract as much of the effort as possible to minimize the military footprint" in Liberia.[8]

The case of OUA illustrates four central elements of present-day military logistics contracting. The first is that the use of contracted support has been institutionalized by the military and is built into operational plans from the beginning. The best example of this is LOGCAP, the primary mechanism through which the U.S. Army procures logistics support. From humble beginnings in 1985, LOGCAP has expanded into a multibillion-dollar program, with LOGCAP contractors providing support for the military across the world, from giant bases in the Middle East to drone facilities in Africa to remote counternarcotics outposts in South America.

Second, and relatedly, contractors support a wide variety of overseas operations, not just wars. The first significant use of LOGCAP, for example, was the UN-sanctioned humanitarian intervention in Somalia in 1992–93. In the late 1990s thousands of local and U.S. contractors were employed to support peacekeeping operations in the Balkans. In recent years contractors have played a critical role in supporting clandestine operations by SOF in Africa, ranging from counterterrorism missions across the Sahel and Maghreb to the campaign against Joseph Kony's Lord's Resistance Army in Central Africa.[9] Like Mary's lamb in the nursery rhyme, wherever the military goes, logistics contractors are sure to follow.

Third, there is little that the military does not outsource to contractors. Consider the list of goods and services provided by contractors for OUA: construction, hazardous material disposal, provision of laundry machines, canvas and tent repair, material handling equipment, maintenance of showers, latrines and sewage, bulk fuel operations, dining facilities, fire prevention, bottled water, power generation, vector (pest) control, water production, and logistics transportation.[10] At the largest bases in the Middle East and Afghanistan troops have access to a wide range of amenities such as gyms, movie theaters, twenty-four-hour food courts and cafeterias, internet and merchandise stores, all provided by thousands of workers recruited from around the world.

It is this last detail—the global composition of the workforce—that is the fourth, and arguably most important, aspect of logistics contracting today. From Kenyans washing clothes in Afghanistan to Bosnians providing vector control in Djibouti to Filipinos building detainment facilities in Guantanamo, U.S. military operations overseas are sustained by a diverse labor pool and global recruiting networks.

None of these four elements are wholly without precedent. But taken together they represent a significant departure from past practices. To substantiate this claim, the rest of this chapter provides an overview and history of logistics contracting by the U.S. military. It asks and answers three questions: What is the scale and scope of contracting in the present day? How does this compare with earlier periods in U.S. history? And how did we get here? I begin with an overview of logistics contracting by the U.S. military from the Revolutionary War to Vietnam. This is followed by a discussion of shifts in contracting priorities and practices in the 1990s, as well as the drivers of these changes. I conclude by outlining the scale and scope of contracting in Iraq and beyond.

Camp Followers, Sutlers, and the Beginnings of Corporate Logistics Contracting

As Christopher Kinsey argues, in contrast to armed mercenaries, "which the state has tried to marginalise . . . since the end of the eighteenth century," logistics suppliers and contractors "have continued to be an important part of the military system."[11] In the Revolutionary War soldiers' needs on both sides were served by a train of accompanying civilians that washed and repaired clothes, cooked food, and sold a variety of goods including liquor, clothing, shoes, tobacco, and soap. These "camp followers" included servants and slaves, wives and children of soldiers, unregulated peddlers, and commissioned sutlers whose trade and prices were prescribed.[12] The British Army also relied on contractors to source and ship food, medical supplies, forage, and coal from England, Canada, and Caribbean colonies to troops in America.[13]

In 1821 sutlers were formally incorporated into the U.S. Army's supply system for frontier posts in the West. This involved appointments that granted exclusive trading rights at an assigned post or with a specific regiment in exchange for submitting to rules and regulations that governed prices and the quantity and types of goods to be supplied. This frontier post supply system, which lasted until the end of the century, could be quite profitable for sutlers, especially when paired with army contracts for other necessities such as lumber, fodder, and freight operations.[14] Transportation contracts were a particularly lucrative source of income, especially following the Mexican-American War (1846–48) that led to the acquisition of vast new tracts of Western territories. By the mid-1850s the military's primary overland transportation contractor in the West, Russell, Majors and Waddell, was making hundreds of thousands of dollars in profits—a vast sum at the time.[15] The sutler system eventually became a source of corruption and scandal as contracts and appointments were often acquired through political patronage and bribes. In 1876, for example, President Ulysses Grant's brother, along with Secretary of War William Belknap, were implicated in a pay-for-posts scheme involving tens of thousands of dollars in bribes.

During the Civil War thousands of sutlers and assorted camp followers played a critical role in providing goods for both Union and Confederate soldiers. They were even used by the Union to supply food and sundries to Confederate prisoners of war.[16] At the same time, sutlers developed a negative reputation among soldiers for price gouging, leading to calls in Congress and state legislatures to abolish the position. The letter of one Kentucky volunteer gives a sense of the disgust that merchants who profited from the war inspired: "Is the word 'scoundrel' exaggerated when applied to such cads? Is it enough to merely give them the name? Should they not become strung up at the closest tree, which is strong enough, to bear them and their heavy sins?" He concluded with the bitter statement that "we soldiers can think only with anger about the money making class."[17]

The money-making class in the Civil War did not just consist of unscrupulous sutlers and camp followers that roused the indignation of ordinary soldiers. Far greater fortunes were realized by contractors that supplied the Union with clothing, blankets, tents, wagons, fodder, weapons, and transportation. Indeed, the war was an economic enterprise on a scale never before seen in the nation's history, with spending by the federal government between 1861 and 1865 exceeding the combined total of all previous U.S. government expenditures.[18] It is important to note that this was a mixed military economy, with significant production and transportation labor performed by public enterprises. The Army's Quartermaster's Department, for instance, "employed over 100,000 civilians, far more than any private American business enterprise of the era."[19] Nonetheless, the majority of goods and services for military operations were procured through contracting.

Following the Civil War, the Army quickly shrank back to prewar levels. When the Spanish-American War began in 1898, it consisted of less than 30,000 troops.[20] The invasions of Cuba and the Philippines, as well as the subsequent counterinsurgency campaign in the latter, and participation in the multinational suppression of the Boxer Rebellion in China in 1900–1, marked a turning point in U.S. military and political history as the country became an overseas colonial power. In the twelve decades since it has maintained a continuous global military presence. However, in 1898 the military—and especially the Army—was ill prepared for overseas operations and heavily dependent on civilian support. It was forced to charter, for instance, eighteen of the twenty ships required to ship the initial expeditionary force and its supplies to the Philippines.[21] It also relied on thousands of Filipino and Chinese laborers for construction projects, and transportation services.[22]

Vastly greater logistics requirements for overseas operations impressed upon civilian and military leaders alike the need for the development of greater support capabilities within the military itself. As secretary of war from 1904 to 1908, former Philippines governor and future president William Howard Taft "recommended the formation of a general service corps to replace civilian employees and soldiers released from line units for duty as wagon masters, teamsters, engineers, firemen, carpenters, blacksmiths, overseers, clerks, and laborers." Reforms in this direction were implemented during his administration and "by the time of World War I it had become generally accepted that enlisted service troops of various kinds should perform most of those duties. Men who had never seen a ship were organized into stevedore battalions, men unfamiliar with motor vehicles were assigned to truck companies, men who had never been near an Army depot were assigned to run them."[23] The view that logistics and service labor was a core military function would persist into the 1990s.

The U.S. military refers to logistics and service tasks performed by uniformed personnel as "organic" support. The development of organic capabilities in the twentieth century produced a remarkable shift in the military's force structure. According to one analysis, in World War I, the European theater in World War II, and the wars in Korea and Vietnam, the percentage of U.S. Army personnel engaged in logistics and life support tasks ranged from 35 percent to 45 percent.[24] A consequence of this growth in uniformed personnel that supported these wars was an, at times, tense relationship between combat forces and logistics and administrative staff, which more often than not operated far from the battlefield. The famed World War II military cartoonist Bill Mauldin captured this dynamic in his book *Up Front*: "It was not enough, the doggies [frontline infantry] felt, to live in unspeakable misery and danger while these 'gumshoe so and sos' worked in the comfort and safety of the city. Hell no. When they came back to try to forget the war for a few days, these 'rear echelon goldbrickers' had to pester them to death. When a man is feeling like this you can't tell him that his tormentors are people like himself, and that they are in the rear because they have been ordered to work there, just as he was ordered to the front."[25] Mauldin was not unsympathetic to those in the rear who, as he notes, were just working where they were ordered to by the military. Others were not so charitable. The phrase "desk jockey" was a popular epithet hurled at those working in the rear in World War II, while in Vietnam "rear echelon motherfuckers" or REMFs were a common target of complaints from combat "grunts."[26]

Even with the dedication of a substantial portion of military personnel to support activities contracting still played an important role in facilitating overseas military operations during the two world wars and various Cold War conflicts. American Expeditionary Forces in World War I negotiated with French authorities to obtain civilian labor and materials necessary for the construction of facilities such as barracks and hospital wards.[27] In the Korean War the military drew heavily on Japanese and Korean labor, employing approximately 100,000 civilians in Korea and 145,000 in Japan.[28] The Army's Japan Logistical Command "estimated that if all the supply and service functions of that command had been carried out without the use of Japanese workers, an additional 200,000 to 250,000 service troops would have been required" to support operations in Korea.[29] One reason for the extensive reliance on civilian labor in this conflict was that the U.S. was also largely responsible for the logistics needs of the South Korean Army and allied UN forces.[30]

By the time of the Vietnam War, the size of organic logistics and life support forces began to come under criticism both within the military and among the U.S. public. In 1967 news reports suggested that only 70,000 of 464,000 uniformed personnel in Vietnam were combat troops. While the Pentagon denied that this estimate was accurate, Secretary of Defense Robert McNamara acknowledged that there was room for improvement in "reducing the ratio of support to combat forces."[31] It was also in Vietnam that U.S. corporations started to play an increasingly visible role supporting troops on the battlefield, with one business publication declaring it a "war by contract."[32]

Contracting was particularly pronounced in the field of military construction, especially during the rapid buildup in 1965–66. At its peak, Pacific Architects and Engineers (PAE), an Army construction and engineering contractor, had more than 21,000 workers in the country.[33] The primary construction contractor was a consortium of four firms—Raymond International, Morrison-Knudsen, Brown & Root, and J.A. Jones Construction—called RMK-BRJ.[34] According to a RMK-BRJ document called "Diary of a Contract," 1966 was

> as wild a period as any human being can imagine. Thousands of people were arriving from the United States, South Korea, the Philippines and 27 other nations; tens of thousands of South Vietnamese were hired and taught construction trade. . . . Not the least of the problems being faced

was building the base for the contractor's own operations—camps, maintenance shops, warehouses, etc. These competed for the labor, materials and time which the soldiers, sailors, airmen marines understandably felt were there to fulfill their own urgent needs. In short it was a period of 20-hour days, 7-day weeks, frayed nerves, deadlines, shortages, and magnificent achievement.[35]

The consortium even published a newspaper in both English and Vietnamese called *Vietnam Builders*, which carried stories on completed construction projects and "human interest" features such as softball games between Filipino workers and U.S. soldiers.[36]

The use of contractors in Vietnam was not without its critics. Brown & Root's contracts were a particular object of contention due to its close ties with President Lyndon Johnson. The company had bankrolled several of his political campaigns, including his successful election to the Senate in 1948. In exchange Johnson helped Brown & Root secure hundreds of millions in federal contracts when he was Senate majority leader and president.[37] In a rather ironic twist, in 1966 Donald Rumsfeld—at the time a U.S. House representative from Illinois—criticized the Johnson administration for poor contracting oversight and reliance on a single contract with RMK-BRJ to provide the bulk of construction support: "Under only one contract, between the U.S. government and this combine [RMK-BRJ] it is officially estimated that obligations will reach at least $900 million by November 1967. . . . Why this huge contract has not been and is not now being adequately audited is beyond me. The potential for waste and profiteering under such a contract is substantial."[38] Four decades later Brown & Root would be called a war profiteer by many within and outside the armed forces, while the administration Rumsfeld worked for would be accused of political cronyism for steering billions of dollars in contracts the company's way.[39]

If one steps back and examines contractor-to-troop ratios—which should be considered rough estimates—from the American Revolution to the First Gulf War in 1990–91, the picture we see is one of fairly consistent but not overwhelming reliance on civilian support, with contractor personnel constituting between 5 and 20 percent of troop levels.[40] There are two exceptions to this trend. The first is the Korean War, which had a 1:2.5 ratio of contractors to troops. As noted above, however, in Korea the military was also responsible for providing the bulk of logistics support for its South Korean and UN allies. Their total numbers were roughly equal to U.S. military forces in the war. When one takes this into account, the ratio of contractors to military personnel in the Korean War is analogous to other wars.

The second apparent anomaly is the First Gulf War, where the contractor to troop ratio is estimated to have been 1:60. This does not mean that civilian labor played a minor role in the conflict. According to one military analysis, outside support "was an essential part of the overall operation. All of the POL [petroleum, oils, and lubricants] and water, most of the construction engineering, most of the port operations, and about 50 percent of the long-haul transportation was provided by External Support."[41] The bulk of these outside support services were sourced and funded by Saudi Arabia and other Gulf countries. Saudi Arabia, for example, "agreed to provide, at no cost to the United States, all fuel, food, water, local transportation, and facilities for all US forces in the Kingdom."[42] Such Host Nation Support (HNS)—and the labor required to provide it—does not appear in contracting calculations because the military did not pay for it. For this reason we will likely never have a good accounting of the military's reliance on civilian workers in the war. However, some of the same Gulf firms that provided support for the military in 1990–91, such as the Saudi Catering & Contracting Company (SCCC), would reemerge as subcontractors in Iraq in the 2000s.[43]

The Transformation of Logistics Contracting

One can trace the emergence of military logistics contracting on the scale that we see today to the years immediately following the end of the Cold War. However, the seeds of change, as Laura Dickinson notes, were planted in Vietnam, with military reports after the war making the case for "continued and increased use of contractors to provide logistical support on the battlefield."[44] It did not hurt that such prescriptions resonated with the "privatization revolution" ushered in by Ronald Reagan's presidency in the 1980s.[45] A key administrative change setting the stage for the dramatic upshift in outsourcing occurred in 1985, when the Army published new regulations establishing LOGCAP to "preplan for the use of civilian contractors to perform selected services in wartime."[46] These regulations were introduced in response to a directive from Congress the previous year to develop contingency contracting capabilities for overseas operations.[47]

LOGCAP contracting remained small-scale for several years, primarily because it was originally designed to be a decentralized program in which various components of an Army command would be responsible for identifying needs and establishing contracts. This led to critiques from officers that LOGCAP was too narrow and limited to functional area support such as transportation of oil supplies.[48] Experience gained in Desert Storm demonstrated the need for more comprehensive support such as the construction and operation of full-service dining facilities. Military planners also increasingly advocated a "turn-key" approach to contracting such as designating a single contractor responsible for the construction of an entire base camp and provision of all support services therein.[49]

In 1992 Secretary of Defense Dick Cheney gave Brown & Root a $3.9 million contract to study how LOGCAP could be reformed to better support soldiers on the battlefield. Brown & Root's report suggested the program be transformed into a single umbrella contract that would provide support services around the world.[50] Another key innovation was the call to fully integrate LOGCAP into planning for possible overseas operations. LOGCAP's prime contractor, the report argued, should be required to develop a worldwide management plan describing "the equipment, personnel and supporting services required to support a force of up to 20,000 troops in 5 base camps for up to 180 days and up to 50,000 troops beyond 180 days."[51] In addition to this, the report recommended that the program's contractor be asked to produce more than a dozen regional plans outlining detailed logistics and engineering support for region-specific planning scenarios used by military commanders. Such plans were necessary because it was envisioned that the contractor would be able to deploy assets within seventy-two hours of initial notice.[52]

Cheney took up Brown & Root's recommendations and put the first LOGCAP contract under this new scheme (LOGCAP I) up for bid. The winning bidder was Brown & Root, a rather curious decision since it was also the firm that wrote the requirements, and the government generally prohibits such arrangements. Four months after the contract was announced the company was tasked with providing logistical support for U.S. forces in Somalia. This was followed by contracts to support humanitarian deployments to Rwanda and Haiti in 1994. LOGCAP was also utilized during military operations in Saudi Arabia and Kuwait that year following a buildup of Iraqi forces along those countries' borders. The following year Brown & Root provided construction and logistics services at Aviano Air Base in Italy in support of the NATO bombing campaign against Bosnian Serb forces. Each of these was a fairly small-scale operation. In total their estimated cost added up to $212 million.[53]

The first time that LOGCAP, and logistics contracting more generally, was utilized on a large scale was the peacekeeping mission in Bosnia starting in December 1995. In the initial year of operations Brown & Root built nineteen base camps for U.S. troops and provided maintenance and logistics support for thirty-two camps in Bosnia, Hungary, and Croatia. In total it earned over $460 million through the LOGCAP contract in that first year alone, more than twice as much as all previous operations combined.[54] In 1997 Brown & Root lost its bid for the new LOGCAP contract (LOGCAP II) to DynCorp. Over the course of this contract DynCorp supported counternarcotics operations in several Central and South American countries, as well as operations in East Timor and the Philippines.

Army officers in Bosnia, however, were generally satisfied with Brown & Root's performance and did not want to switch contractors midstream.[55] The firm was therefore awarded a separate "Balkans sustainment contract." This contract was subsequently extended to Kosovo when the UN mission was established there in 1999. In a short time roughly 5,000 Kosovars and Macedonians were working for Brown & Root, making it the largest employer in the region.[56] Such a large workforce was needed because there was little that the company was not asked to do. As one Army officer observed at the time, "When soldiers first step off airplanes in Kosovo, they are met not by their commander, but by a Brown and Root civilian worker who tells them where they can pick up their gear and assigns them to their barracks."[57] In the end the company made over $2 billion from the Balkans sustainment contract.[58]

Though LOGCAP was developed by the Army, the Navy and Air Force were allowed to utilize the program if needed, as they did for the 1995 operations at Aviano. Both services also developed their own contingency contracting schemes modeled on the program, beginning with the Navy's Construction Capabilities (CONCAP) program in 1995, which was followed by the Air Force Contract Augmentation Program (AFCAP) in 1997. Perini won the first CONCAP contract, but lost the rebid in 2001 to Brown & Root. The following year the program was utilized for the construction of expanded detainment facilities in Guantanamo with—as discussed in chapter 3—a Brown & Root subcontractor flying in hundreds of Filipino construction workers for the project.[59] The winning bidder for the first AFCAP contract, Readiness Management Support LC, was called upon to perform a variety of tasks worth $170 million in the first five years of the program, including constructing refugee camps in Kosovo, refurbishing airfields in Ecuador used for counternarcotics operations, reconstructing damaged infrastructure in Guam following Typhoon Paka in 1997, and design work at Ali Al Salem Air Base in Kuwait.[60]

In short, by the early 2000s logistics contracting for overseas operations was well established. Thus even before the U.S. invaded Afghanistan and Iraq there was little question that contractors would play a central role in supporting future U.S. military operations. Before moving on to this part of our story, though, it is necessary to consider why contracting expanded so rapidly in the 1990s.

There are several reasons for this transformation. One of the most important, as noted above, was the rise of privatization, which was nourished by free market proponents in the Reagan administration. In 1982 the president commissioned an investigation into federal government inefficiency constituted by a committee of private sector executives. The resulting report, presented two years later, advocated greater outsourcing of government provided services. The commission's chair, chemical and materials industrialist J. Peter Grace, also published an accompanying book extolling privatization.[61] By the early 1990s the supposed benefits of privatization had become a bipartisan mantra, as exemplified by the Clinton administration's "reinventing government" commission, which also advocated outsourcing a wide range of government activities.[62] Not to be outdone, the Department of Defense (DoD) followed with a 1996 report on "outsourcing and privatization" by the influential Defense Science Board Task Force that that concluded that "all DoD support functions should be contracted out to private vendors except those functions which are inherently governmental, are directly involved in war fighting, or for which no adequate private sector capability exists or can be expected to be established."[63]

Privatization fever in the military spread well beyond logistics support functions. By the end of the 1990s large swaths of military intelligence gathering and analysis were also being performed by contractors, a trend that has only accelerated since 9/11. This despite the traditional view of such intelligence tasks as core national security competencies.[64] Since the 1990s the military has also moved to privatize a range of social welfare provisions that it provides to uniformed personnel, from housing and health care to family support services and recreational programs.[65] Like the military, DoS has also been radically transformed by outsourcing. Its development branch, USAID, experienced a 45 percent cut in employees between 1980 and 2001, turning the agency into, in the words of one analysis, a "check writer to contractors."[66] Moreover, DoS has become one of the largest consumers of private security industry services, employing thousands of armed contractors to protect diplomats and embassies around the world.[67]

As Maya Eichler demonstrates, another factor driving the privatization of military labor in the U.S. was the termination of male conscription (the draft) in 1973. The central element of her argument is that the "introduction of the all-volunteer force redefined military service as a market relation, even as the citizen-soldier is still invoked and symbolically significant." Severing the linking between citizenship and military service is consequential because if "citizens are no longer required to participate in the public provision of security, the outsourcing of military work becomes justifiable to a much larger extent."[68] Not coincidentally, some of the most prominent opponents of the draft were "Chicago School" economists like Milton Friedman.

Another catalyst driving outsourcing was the dramatic downsizing of the military following the end of the Cold War. In 1987 there were 2.2 million active-duty and full-time guard and reserve troops. By the end of the Clinton administration this number had fallen below 1.5 million, a more than 30 percent reduction in forces.[69] These cuts fell disproportionately on service and support components of the military. The U.S. Army Materiel Command (AMC), for instance, experienced a 60 percent reduction in personnel.[70] At the same time the demand for military operations rose dramatically in the post-Cold War period.[71] In response the military became intensely concerned with improving the "tooth to tail" ratio—that is, outsourcing noncombat support functions so that a greater percentage of remaining troops are engaged in combat activities.[72] As Secretary of Defense William Cohen put it in 1997, "We can sustain the shooters and reduce the supporters—we can keep the tooth, but cut the tail."[73] Cohen's successor, Donald Rumsfeld, echoed this sentiment in 2003 when he stated that "something in the neighborhood of 300,000 men and women in uniform are doing jobs that aren't for men and women in uniform."[74] Two years later he would proudly tell Congress that "mostly administrative and facilities related" duties performed by contractors were "freeing up additional tens of thousands of military personnel for military responsibilities."[75]

A smaller, more focused military also has operational and political advantages. For example, when force caps are imposed for a particular mission, civilian contractors do not count toward troop limits. In 1995 the military estimated that it needed a force of 38,000 troops to fulfill peacekeeping duties in the Balkans. Congress set a ceiling of 25,000 (20,000 in Bosnia and 5,000 in Croatia), with no more than 4,300 of this number permitted to be reservists. The Army was able to manage these restrictions because it could turn to Brown & Root to provide needed logistics support without using military personnel. Indeed, the chief operations planner for the Bosnian peacekeeping mission bluntly concluded that "the truth of the matter is that we are in a force cap environment, be it Army end strength, or operational deployment. Therefore, I think it is reasonable to believe that LOGCAP or some form of contractor support will always be with us and that it is therefore something that always should be built into the plan."[76] Five years later the Clinton administration asked Congress for money for a counternarcotics initiative in South America called "Plan Colombia." Congress approved the funding, but with a stipulation that no more than 500 troops could be deployed to support the operation. To make up capacity gaps, 300 military contractors were permitted to be hired.[77]

Even when overseas operations do not have mandated force caps, such as the wars in Iraq and Afghanistan, the extensive use of contractors for logistical support lowers the number of military personnel that need to be deployed. This is especially important when it comes to the politically sensitive activation and deployment of reservists, which have constituted a disproportionately large percentage of organic support and service units since reforms introduced at the end of the Vietnam War.[78] Hundreds of thousands of reservists were called up in the first three years of the Iraq War, stretching reserve forces to the breaking point and disrupting families and communities across the country.[79] Without the extensive use of contractors the reserve system would likely have been overwhelmed.

A final factor to consider is that contractor deaths are far less politically salient—even when they are U.S. citizens—than uniformed personnel. This "disposable army" does not come home in flag-covered coffins, and is rarely mentioned when discussing casualties and the human costs of war.[80] As Deborah Avant and Lee Sigelman observed in 2010, "Military casualty figures are routinely collected and released. The names and faces of military casualties in Iraq and Afghanistan are shown nightly on *The PBS News Hour*. Coverage of military deployments is virtually automatic. There is no such coordinated or automatic diffusion of information about contractors, nor are there triggers to alert the media. Casualty figures routinely collected and released by the military exclude contract personnel, thus reducing information about the human costs of war."[81] Contractors tend to be treated, in other words, as another category of uncounted civilian bodies in the various U.S. wars.[82]

Logistics Contracting in Iraq and Beyond

The scale of U.S. military logistics contracting since the early 2000s is remarkable. In 2001 Brown & Root, by then called KBR after a merger with the British engineering and construction firm M.W. Kellogg in the late 1990s, won the bidding for the third iteration of the LOGCAP contract (LOGCAP III). In total the company would be paid more than $40 billion during the life of this contract (2001–8), primarily for logistics support in the wars in Iraq and Afghanistan.[83] While LOGCAP was the largest contracting program—and KBR the largest contractor—during this period, it was just one of many mechanisms for contracting logistics services. Sourcing food supplies and delivering them to bases in the region, for instance, was the responsibility of the Defense Logistics Agency (DLA). In 2003 DLA contracted out the job of supplying food to troops in Iraq and Kuwait to a politically connected Kuwaiti firm called Public Warehousing Corporation (PWC).[84] In the first four years of operations it earned more than $6 billion.[85] In 2009 PWC (now called Agility), was charged with fraud and received a three-year suspension from receiving further federal contracts.[86] KBR and its various subcontractors were also the subjects of numerous claims of cost overruns and fraud.[87]

At this point it is useful to briefly introduce the types of contracts used for contracting during overseas operations, which can be divided into two broad categories: "fixed price" and "cost reimbursement" contracts. With the former the government and contractor agree on a price for clearly specified services or goods.

The contractor assumes the risk for cost overruns, and its profit is determined by the difference between the price and the cost of delivering the goods or services. In circumstances where it is difficult to precisely specify services or determine costs beforehand—such as operations in warzones—cost reimbursement contracts are more common. With cost reimbursement contracts a price estimate is settled on, but the government assumes the risk of "reasonable" cost overruns due to contingencies. Cost reimbursement contracts usually include fees that determine a contractor's profits based on a percentage of estimated costs, a fixed amount, and/or performance incentives.[88]

PWC's food delivery contract with DLA included reimbursement for the purchase price of food from suppliers and a fixed price "distribution fee" for transportation of the supplies from the U.S. to bases in Iraq and Kuwait. PWC was accused of negotiating price discounts with favored suppliers, but not passing these savings on to the government as required, thus increasing its profits.[89] LOGCAP III was a cost reimbursement contract, with a base fee of 1 percent of estimated costs and an award fee of up to 2 percent based on performance incentives.[90] Former LOGCAP III manager Charles Smith notes that this type of contract is frequently misunderstood, with most of KBR's critics claiming that the company would increase costs of services to increase its fees. However, if the actual costs of work exceeded the estimate this would not produce extra fees for KBR, and it could even lower performance incentives. Instead, Smith observes, "the problem was the period between starting work and negotiating the fee cost base."[91] He is referring to the fact that the way LOGCAP typically worked in Iraq is that KBR would be issued an unpriced task order—or a modification to an existing order—begin work, and then negotiate the estimated cost with the government. This gave it a strong incentive to increase costs at the beginning of the contract, with the hope that they would be accepted into the final estimated cost base, thus increasing its fees. According to Smith, KBR pervasively submitted inflated cost estimates that it could not substantiate, but military superiors undercut his attempts to hold the company accountable, eventually leading to his dismissal. These charges are corroborated by Defense Contract Audit Agency audits that found that "DoD contracting officials rarely challenged" KBR's cost estimates even though these estimates were "later found to be greatly inflated."[92]

In addition to fraud and profiteering, logistics contractors in Iraq and Afghanistan have also been accused of mistreating workers, beginning with a series of news articles in 2004 that highlighted problems of trafficking and other labor abuses, including wage theft and substandard living conditions.[93] One of the more astonishing details to come out of these articles was that the military had no idea just how many people were working for it. In the rush to build bases and implement various reconstruction projects following the invasion, it entered into an

untold number of contracts with international and Iraqi firms. The largest contractors like KBR in turn outsourced their tasks to a panoply of subcontractors.
Responding to criticism that its contracting oversight in Iraq was lacking, CENT-
COM conducted an initial contractor census in December 2006. According to its
calculations there were roughly 100,000 contractors in the country, not including
subcontractors, which it was unable to estimate.[94] Over the next two years it refined and expanded its census efforts, culminating in the publication of the first
command-wide census in August 2008. As discussed in the introduction, these
quarterly censuses offer a useful aggregate picture of military contracting in the
region over the past decade. But it is possible to gain a more granular understanding by examining raw data from censuses conducted in Iraq between 3rd quarter
2007 and 2nd quarter 2008 that the military released following FOIA requests by
journalists.[95] To date these data have not been subject to analysis, despite the fascinating window into contracting that they offer.

The first thing one notices is that the censuses were a work in progress. Reporting procedures for data columns such as "mission" evolved over the course
of the year, from at times detailed descriptions in the first census—"SST [shit-
sucking-truck] services. Fuel is provided for one truck located onsite. The other
four trucks are fueled outside by the subcontractor" for a waste removal contract—
toward more standardized and anodyne descriptions like "base support" by the
last one. It also appears that the military was still having difficulties in accurately
tracking subcontracted labor, especially in cases of pyramid subcontracting where
a subcontractor in turn subcontracts out responsibilities to another firm. The first
census lists 2,109 contracts and a workforce totaling 136,655. By the last census
these figures had risen to 2,452 and 149,378, respectively. The jump in contracts
and workers was driven in part by the troop surge taking place during this same
period, but it also appears that a portion of this increase reflects better reporting.

The final Iraq census in 2nd quarter 2008 is not only the most refined, it also
offers a snapshot of contracting near its peak later that year. During this period
the military estimated that 20 percent of the contracting workforce were U.S. citizens, 38 percent TCNs, and 42 percent LNs. The relatively large percentage of
Iraqi workers reflected both an increase in employment for training and reconstruction projects connected to the surge, and efforts by military commanders to
direct more small logistical support contracts toward Iraqi businesses following
the introduction of the "Iraqi First" program in 2006.[96] The majority of contracting was conducted through either LOGCAP or the ad hoc, theater-based contracting framework, Joint Contracting Command-Iraq/Afghanistan (JCC-I/A),
with the former accounting for 37 percent of workers and the latter 35 percent.
The JCC-I/A data are the least systematic with regard to categorization and
description of contracts. This appears to be due to its decentralized structure,

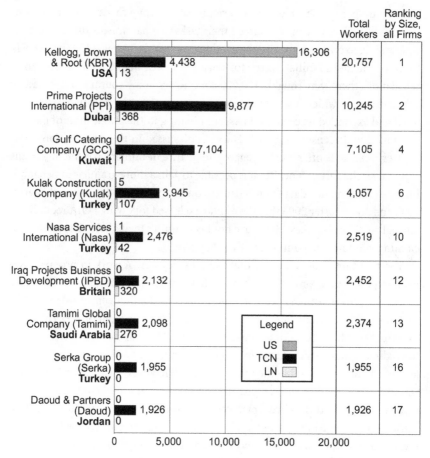

FIGURE 2.1. Largest LOGCAP firms by number of workers and ranking among all PMCs in Iraq, 2nd quarter 2008

with twelve regional contracting offices operating across the country in 2008. Nonetheless, looking at the data it is possible to discern that JCC-I/A contracts fell under three main areas: 1) private security, 2) reconstruction and training, and 3) logistics support services, with an emphasis on hiring Iraqi companies and laborers.

LOGCAP was not only the most significant contracting program in Iraq, the companies that received contracts under its umbrella were also many of the largest military contractors in the country. Leading the way was the prime contractor, KBR, which had nearly 21,000 employees, including thousands from Southeast Europe, which composed the bulk of its TCN workforce. Following KBR was its largest subcontractor, Prime Projects International (PPI), with more than 10,000 workers. In total nine of the twenty largest PMCs by workforce in

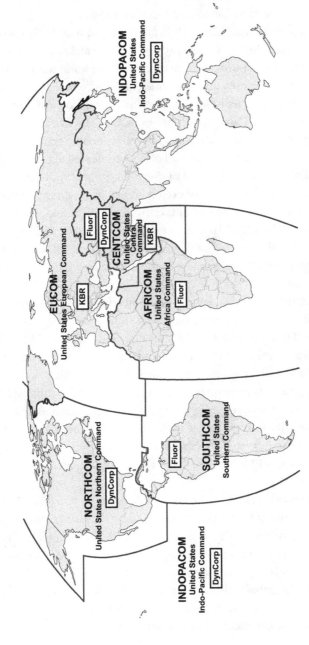

FIGURE 2.2. U.S. military geographic combatant commands and LOGCAP IV contractors

Iraq (KBR and eight of its subcontractors) were firms that received all or a significant portion of their contracts through the LOGCAP program (figure 2.1).[97] The largest non-LOGCAP contractor, with 8,795 workers, was L-3 Communications, which provided Arabic interpreters for the military.

Combined, these nine LOGCAP firms employed over 53,000 people, with 68 percent of them being TCNs, mostly from South and Southeast Asian countries. The prevalence of workers from this part of the world was due to the provenance of KBR's subcontracting companies, nearly all of which were based in the Middle East. Following well-established recruiting practices and pathways, firms like PPI (Dubai), Gulf Catering Company, or GCC (Kuwait), and Serka (Turkey) tapped recruiting brokers and agencies in countries like India, the Philippines, Nepal, and Sri Lanka to amass the pool of laborers needed to fulfill their growing contractual obligations in Iraq.[98] As I discuss in subsequent chapters, they also often brought with them a host of exploitative labor practices that parallel conditions experienced by labor migrants elsewhere in the region. Ironically, KBR's 2001 bid for the LOGCAP III contract proposed that the company would self-perform most of the required work, as it did in Bosnia, rather than relying upon subcontractors, with the government considering this a positive aspect of the proposal that would lessen risk.[99]

In part in response to accusations of profiteering and mistreatment of workers, the military decided to move away from a single prime contractor approach with its next LOGCAP award. In 2008 the new contract (LOGCAP IV) was split among three companies (KBR, Fluor, and DynCorp), who compete for task orders.[100] Similar to the Balkans sustainment contract, KBR was chosen to continue supporting military operations in Iraq until the withdrawal of troops in 2011. However, its operations in Afghanistan were turned over to Fluor and DynCorp, with the former charged with providing support to bases in the northern half of the country and the latter the southern. Contracting in Afghanistan peaked in spring 2012 with more than 117,000 contractors supporting military operations.[101]

In addition to moving away from a single prime contractor, a second innovation coming out of the LOGCAP IV contract is the further integration of outsourcing with the military's worldwide force posture. Five of its six geographic combatant commands are now assigned to a prime LOGCAP contractor, while KBR, Fluor, and DynCorp continue to split task orders in the sixth, CENTCOM (figure 2.2).[102] A global logistics contracting network and workforce for a global military presence. We now turn to the histories that explain the prevalence of Filipino and Bosnian laborers in this workforce.

COLONIAL LEGACIES AND LABOR EXPORT

No other nation has felt the force of American power so closely, so constantly throughout Washington's century-long rise to world leadership. No other nation can reveal so much about the character of America's international influence, both direct colonial rule and diffuse global hegemony.

—Alfred McCoy

It is the rare news story on foreign labor at overseas U.S. bases that does not mention workers from the Philippines. From Patap Chatterjee's description of "coffee shops run efficiently and politely by Indian and Filipino migrant workers, who serve up espresso chai latte and mocha frappers" at massive bases in the Middle East, to Sarah Stillman's observation about Filipinos who "launder soldiers' uniforms" in Afghanistan, to David Phinney's discussion of labor strikes in Iraq, Filipino workers are ubiquitous.[1] These accounts reflect the fact that many of the largest firms providing logistics support for the military over the past fifteen years have relied heavily on Filipino labor. Unfortunately CENTCOM censuses do not provide information on the country of origin of TCNs, so it is not possible to determine the number of Filipinos who have worked in Iraq and Afghanistan, but by all indications the Philippines has been one of the most significant sources of labor. Even after the country imposed travel bans to Iraq and Afghanistan, thousands of Filipinos were added to the payrolls of military contractors in the region.

Why is the Philippines one of the primary suppliers of labor for the U.S. military? Certainly cost is a consideration. KBR's largest subcontractor, PPI, for example, typically paid its Filipino workers in Iraq around $500–600 a month. But cheap labor is just one element. Additionally, interviews I conducted in the Philippines, along with various news accounts, suggest that South Asian workers have tended to be paid less than Filipinos, especially when one takes into account the exorbitant recruiting fees they must pay back after receiving a job—fees that are often split between recruiters and subcontractors.[2] Thus to fully understand the link between current U.S. military operations and Filipino labor requires a more historical perspective.

In this chapter I describe two historical formations that influence this relationship. The first followed the U.S.'s annexation of the Philippines following the Spanish-American War of 1898. During the subsequent colonial period, and in the decades following independence when the Philippines operated as a U.S. client state, Filipino labor was enrolled to facilitate a number of military and civilian projects. The second was the emergence of labor export as a development strategy pursued by the Philippines in the 1970s. Part of a broader labor import-export regime between newly wealthy Gulf states and poor South and Southeast Asian countries that materialized that decade, this assemblage—along with its associated practices and labor flows—has become imbricated with U.S. military logistics outsourcing through the extensive reliance on subcontracting companies from the Middle East. After examining these two formations I conclude the chapter by explaining how the prevalence of Filipino labor in the Middle East, and the Philippines' unique historical relationship with the U.S., shaped President Arroyo's decision to support the invasion in Iraq, with an eye to the economic and political benefits she anticipated would accrue.

Making the Empire Work

The year 1898 is a momentous one in the history of both the U.S. and the Philippines, the year in which these two countries' histories became inextricably joined through the former's defeat of Spain and annexation of several of its colonies, including the Philippines. For many scholars it also represents a transformative moment, the point when the U.S. transitioned from a republic to a global imperial power. This transformative view was widely held by contemporaries—both proponents and opponents of the new colonial territories. "By the acquisitions made during this period, the United States has definitively entered the class of nations holding and governing over-sea colonial possessions" observed William Willoughby, an economist appointed to an administrative position in Puerto Rico in 1901.[3] Rudyard Kipling's famous poem, "The White Man's Burden" (1899) exhorted the U.S. to embrace colonial rule over the Philippine islands.[4] Supreme Court justice John Marshall Harlan, a staunch critic, deemed colonial administration of the new territories "a radical and mischievous change" and argued that "the idea that this country may acquire territories anywhere upon earth, by conquest or treaty, and hold them as mere colonies or provinces—the people inhabiting them to enjoy only such rights as Congress chooses to accord them—is wholly inconsistent with the spirit and genius, as well as with the words, of the

Constitution."[5] Not all subscribe to this vision of 1898 as rupture. The historian Paul Kennedy asserts that "from the time the first settlers arrived in Virginia from England and started moving westward, this was an imperial nation, a conquering nation."[6] Increasingly, scholars are adopting this longer perspective, arguing that nineteenth-century campaigns to exterminate and/or displace Native Americans and the annexation and settlement of vast territories across the American West need also be situated in the context of U.S. imperial expansion.

Regardless of one's take on this debate, one fact is clear: from the beginning of colonial rule in the Philippines, as well as in the decades following independence in 1946, Filipino labor has played a crucial role in the spread of U.S. empire across the Pacific and beyond. As early as 1901, when the Philippine insurgency against their new colonial overloads was still raging, the Hawaiian Sugar Planters Association (HSPA) began inquiring about the possibility of importing Filipino labor to work on its members' sugar plantations.[7] It directed its inquiry to the Bureau of Insular Affairs, the newly created bureaucratic entity under the auspices of the Department of War that was charged with overseeing the administration of U.S. overseas possessions. Founded in the aftermath of the 1893 overthrow of Queen Liliuokalani, ruler of the Kingdom of Hawaii, HSPA was itself a significant node in the new network of U.S. territorial possessions. Many of its members had been key figures behind the coup and subsequently became proponents of annexation, a goal achieved when Hawaii was incorporated as a U.S. territory in 1898.

Assembling Filipino labor for Hawaiian sugar plantations foreshadowed the current labor export system in the Philippines. In 1915 a Philippine Bureau of Labor was established to regulate labor recruitment, supplanting the Bureau of Insular Affairs. HSPA recruiting agents were required to obtain permits from the Bureau of Labor to set up offices in designated provinces. Signed labor contracts were mandated, and in 1915 labor agreements began including free transport back from Hawaii after the completion of a three-year contract. Mechanisms to monitor labor conditions were also developed, most notably the position of resident labor commissioner based in Honolulu, which was established in 1923 in response to petitions by Filipino plantation workers. Ultimately this did little to improve the conditions of workers as the commissioner, Cayetano Ligot, typically supported plantation owners in labor disputes. This is not surprising since "Filipino government leaders remained under the ultimate supervision and control of the United States" and were "expected to act according to U.S. interests."[8] Nonetheless, HSPA agents did not struggle to find willing labor migrants. By 1922, 41 percent of the plantation workforce was Filipino.[9] In total more than 125,000 Filipinos had traveled to Hawaii to work on the sugar plantations by 1946.[10]

In the introduction to their edited volume on labor and early U.S. empire, *Making the Empire Work*, Daniel Bender and Jana Lipman claim that "perhaps the most obvious cohort of workers who built the U.S. empire are those who labored in agriculture."[11] Arguably no group filled this need for cheap, flexible, agricultural labor more than Filipinos in the first half of the twentieth century. In addition to Hawaiian sugar plantations, by the 1930s tens of thousands of Filipino laborers could be found at farms across California, harvesting lettuce in Watsonville, melons in the Imperial Valley, and fruit trees in the Central Valley. They were also recruited to work in apple and cherry orchards in Washington and the hop fields of Oregon. Several thousand more were employed in the burgeoning canned salmon industry in Alaska, and by the 1920s the canneries were "a regular stop on the seasonal labor circuit that stretched from Southern California to Alaska."[12]

The demand for Filipino labor was tied to two political and legal developments. The first was a series of exclusionary immigration policies that closed off other labor flows, beginning with the Chinese Exclusion Act (1882), followed by an informal "gentleman's agreement" (1907) limiting Japanese immigration and the creation of an "Asiatic Barred Zone" (1917), and culminating in the Immigration Act of 1924, which sharply curtailed immigration outside of Northern Europe. The second concerns a series of opinions issued by the Supreme Court in 1901 in the wake of the occupation of Puerto Rico and the Philippines. Known as the "Insular Cases," these opinions determined that these U.S. overseas colonies should be considered "unincorporated territories," a new legal category that situated them "in a liminal space both inside and outside the boundaries of the Constitution, both 'belonging to' but 'not a part' of the United States."[13] Filipinos, consequently, were classified as U.S. nationals—but not citizens—and thus were not bound by the immigration laws passed by Congress. JoAnna Poblete usefully categorizes those suspended in this liminal and subordinate political and legal status as "U.S. colonials." Exempt from immigration restrictions during this period, U.S. colonials like Filipinos and Puerto Ricans experienced remarkable labor mobility. In addition to travel to the mainland, this often involved *intracolonial* movement "from a colonized home region to another colonized location," as was the case with Filipino sugar plantation workers in Hawaii.[14]

While their status as U.S. nationals facilitated mobility, it did not protect Filipino migrants from the experience of racism and exploitative labor conditions. This was especially the case with the agricultural industry in California where they competed with poor whites in the labor market, particularly following the migration of tens of thousands of destitute families to California following the Dust Bowl in the Plains. Filipinos also threatened entrenched racial hierarchies, especially men who dated or married white women. Thus in the eyes of many white Americans at the time they represented a "foreign invasion that challenged Americanness, as non-whites they were a threat to whiteness, and as a mobile workforce who did not need U.S. passports they were regular competitors for employment."[15] In 1930 violent riots targeting Filipinos spread across California. Growing opposition to their presence culminated in the Tydings-McDuffie Act (1934), which reclassified Filipinos as aliens, established strict limits on immigration—with the exception of the territory of Hawaii, "based on the needs of [sugar] industries"—and laid out a ten-year process for Philippine independence.

A second—and more significant for this story—labor cohort involves Filipinos who worked for the U.S. military. As with agricultural work, this can be traced to the beginnings of colonial rule. One of the first major infrastructure projects pursued by the new colonial authorities was construction of a mountain retreat in Baguio conceived along the lines of hill stations built by the British in India. Nearly 5,000 feet up the Cordillera mountain range in central Luzon, the Baguio retreat, and especially the road leading to it, was a labor-intensive undertaking. One of the biggest challenges facing military officials was recruiting and retaining workers given the low pay offered, dismal living conditions and the backbreaking and dangerous nature of the work.[16] In 1901 an army officer overseeing construction of the road complained that "securing native laborers continues to be a most serious difficulty." To overcome this problem the military experimented with the use of prison labor in 1903, a short-lived scheme that "ultimately proved costly and accomplished nothing in the way of road building."[17] Baguio is but one of many examples of the use of Filipino labor in support of U.S. military objectives during the colonial period. Another was the Philippine Scouts, America's colonial army version of the Gurkhas.[18] Looking for a more effective means of countering the ongoing insurgency, the U.S. Army incorporated the Scouts in 1901. Initially a 5,000-strong force, they formed the backbone of U.S. forces in the Philippines up to World War II.

More widespread and enduring was the use of Filipino labor by the U.S. Navy, which began in 1901. By the 1920s Filipinos constituted roughly 5 percent of the Navy's workforce, serving predominately as stewards who cooked, washed dishes, and cleaned officers' quarters.[19] Reflecting racial and colonial hierarchies that prevailed, in subsequent years they often served alongside African Americans, who were also relegated to these positions.[20] Even independence did not end this labor arrangement. As part of the 1947 Military Bases Agreement concluded between the Philippines and U.S., the Navy was given the right to continue enlisting Filipinos. In fact, following President Truman's 1948 executive order to desegregate the armed forces, demand for Filipino stewards grew as African Americans began experiencing opportunities to rise up the ranks. By the 1960s the Navy was receiving as many as 100,000 applications from the Philippines a year. In addition to pay rates that exceeded most salaries in the Philippines, the prospect of

U.S. citizenship after three years of service was a strong inducement. Yet Filipinos' subjugate status as a "brown skinned servant force" continued.[21] In 1970 a scathing article in the *Washington Monthly* characterized the Navy's Filipino recruitment program as "a remnant of colonial rule." Pointing out that over 80 percent of the nearly 17,000 Filipino citizens serving in the Navy worked as stewards, and that Filipinos constituted more than 80 percent of the steward class personnel in the service, the article concluded that the Navy was in effect operating as a "floating plantation" that used Filipinos as an "unending source of docile, cheap, and unquestioning labor."[22]

Another remnant of colonial rule was the continued use of several bases by the U.S. military, most notably Subic Bay Naval Base and Clark Air Base, as stipulated by the Military Bases Agreement. One of the more controversial aspects of this agreement in the Philippines—and that which demonstrated most clearly the country's subordinate status as a former colony and now pliant client state—was language that gave the military exclusive jurisdiction over any criminal offenses committed on bases, even those involving Filipino nationals. This grant of extraterritorial jurisdiction contrasted sharply with the 1951 NATO Status of Forces Agreement (SOFA) that expressly prohibited U.S. forces in Europe from exercising jurisdiction "over persons who are nationals of or ordinarily resident in the receiving State."[23] Following years of protests, an amended agreement in 1965 brought jurisdictional language in line with the NATO SOFA. Yet opposition to U.S. bases continued, reaching a crescendo in the 1980s due to their association with the repressive Ferdinand Marcos regime.[24] This opposition was tempered in part by the substantial role that the bases played in the Philippine economy. In 1987 more than 42,000 Filipinos worked on U.S. bases, earning salaries significantly higher than local prevailing rates. With total wages reaching $82 million, this represented "the second largest payroll in the Philippines, topped only by that of the government itself."[25] Nonetheless, following Marcos's overthrow in the People Power Revolution of 1986, several years of tense negotiations culminated in a decision by the Philippine government to close the bases in 1992.

The presence of Filipino labor on military bases was not confined to the Philippines during the Cold War. In fact it was during this period that the practice of recruiting Filipinos to work at overseas bases flourished. From the late 1940s to the early 1960s thousands could be found on the occupied island of Okinawa, supporting the military construction boom, working as cooks and performing various administrative tasks.[26] And during this period tens of thousands of Filipinos were recruited to work for the military on Guam, Wake, and other island territories across the Pacific.[27] According to military officials, Filipinos also began working at Guantanamo after the Cuban Revolution in 1959 cut off local labor flows.[28] And as noted in chapter 2, Filipino engineers and construction laborers

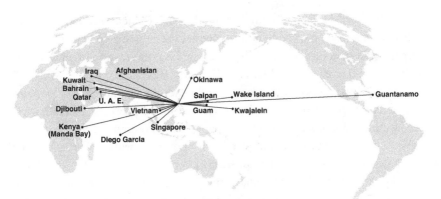

FIGURE 3.1. Filipino labor flows to U.S. military bases from the end of World War II to present

were widely employed by RMK-BRJ and other contractors in Vietnam. By 1980 Filipino laborers could be found as far afield as Diego Garcia.[29] Several years later they were employed by the U.S. contractor Burns and Roe, which managed facilities for the U.S. Navy in Singapore.[30] Indeed, if U.S. empire is defined by its global network of bases, as a number of scholars have argued, then tracing the flows of Filipino labor that make this empire work provides a remarkable—if partial—mapping of critical past and present nodes in the network (figure 3.1).[31] This is especially true of island bases, whether those located on unincorporated territories or sites such as Guantanamo and Diego Garcia, where the "ligatures between colonialism, violence and the law" have produced exceptional regimes of sovereign authority and jurisdiction.[32]

Exporting Labor

The Philippines has a long history of integration with the world economy. In the late sixteenth century Manila was the epicenter of a Spanish-Pacific trade network, facilitated by regular visits from Chinese junks and the development of the "Manila galleon" route to Acapulco.[33] Additionally, as detailed above, from the beginning of U.S. colonial rule Filipino labor was a desired commodity for both military and civilian projects on the mainland and overseas territories. So in one sense the emergence of labor export as a development strategy represents a continuation and deepening of the country's participation in global economic circulation. Yet it also, as Robyn Magalit Rodriguez points out, constitutes a striking example of government-promoted and -regulated "labor brokerage."[34]

This policy was formally implemented in 1974 when Ferdinand Marcos introduced a new labor code by presidential decree. This code provided the institutional structure for labor export, including the creation of an Overseas Employment Development Board (OEDB) and National Seamen Board (NSB). The OEDB was charged with promoting the overseas employment of Filipinos (whom Philippine agencies refer to as Overseas Filipino Workers [OFWs]), overseeing the conditions and terms of employment on a "government-to-government basis," and the recruitment and placement of overseas workers for land-based employment, with similar duties for sea-based workers discharged by the NSB.[35] Initially the state envisioned a phase-out of private recruiting agencies, which predated Marcos's presidential decree and were to be replaced by government monopoly, but this plan was reversed in 1978 due to intense lobbying by the agencies and the fact that it had become clear by then that the government was incapable of responding to dynamic global labor markets as nimbly as the agencies.[36] In 1982 the OEDB, NSB, and other administrative offices were merged into a new Philippine Overseas Employment Administration (POEA), which is currently the primary entity involved in the regulation of overseas employment. Among the POEA's responsibilities are licensing recruiting agencies, regulating the recruiting process, marketing Filipino labor to other countries, approving manpower requests from foreign employers, and monitoring overseas labor and political conditions.

Demonstrating the central importance that the Philippines places on labor brokerage as a means of raising capital, the code also called for mandatory remittances of foreign exchange earnings by labor migrants.[37] In this regard labor export has been a great success, with both the number of OWFs and amount of remittances exploding over the ensuing four decades. According to government estimates, in 2015 there were roughly 2.4 million OFWs working abroad.[38] The following year remittances from OFWs hit a record high of $26.9 billion, which represented nearly 10 percent of the country's total GDP.[39] Indeed, President Arroyo was scarcely exaggerating when in 2006 she called overseas Filipino workers "our greatest export" and "the backbone of the new global workforce."[40] Her rhetoric echoed that of her two immediate predecessors, Joseph Estrada and Fidel Ramos, who lauded OFWs as "economic saviors" and a "major pillar of national development," respectively.[41] This last claim is more debatable as there is little evidence that labor export constitutes a viable development strategy. In fact, since the 1970s the Philippines has grown at a substantially slower pace than other emerging economies in the region such as Thailand, Indonesia, China, Malaysia, and Vietnam.

The primary driver of labor export for the Philippines has been the insatiable demand for labor by oil-rich countries in the Middle East. In 1975 just 1,500 overseas workers (representing 12 percent of total OFWs) were deployed to the Middle East. By 1983 more than 300,000 Filipinos (representing more than 80 percent of total OFWs) were working in the region.[42] Beginning in the late 1980s East and Southeast Asia also emerged as a substantial destination as newly industrializing countries in the region, especially Taiwan, South Korea, and Malaysia, turned to foreign workers to make up labor shortfalls in their export-oriented manufacturing enterprises. Filipinas were increasingly recruited for domestic work in the region as well. With the relative decline of this labor flow since the early 2000s, the Middle East has regained its position as the predominate destination for Filipino labor. In 2015 more than half of the 2.4 million OFWs worked in one of four Gulf states (Saudi Arabia, the United Arab Emirates [UAE], Kuwait, and Qatar).[43]

As massive as Filipino migration to the Middle East is, it represents just a portion of the broader labor import-export assemblage linking workers from poor South and Southeast Asian countries with Gulf petro-states. In 1985 the foreign workforce in the six Gulf countries exceeded 5 million, with laborers from India and Pakistan providing the largest contingents of labor.[44] Twenty years later, an estimated 17 million foreigners worked in the region.[45] There are few jobs that this massive workforce does not do, from highly remunerated occupations like banking and petroleum engineering—typically conducted by European or U.S. "expats"—to low-paid jobs in retail, construction, and domestic care that most workers from Asia are recruited to perform.

This labor import-export dynamic is critical for understanding the prevalence of military workers from South and Southeast Asian countries like the Philippines, India, Pakistan, Sri Lanka and Nepal. The reason for this is that many of the largest military subcontractors are based in the Gulf states. When these subcontractors look for labor, they turn to long-standing and well-established recruiting pathways and firms. One example of this is Dubai-based PPI, which employed nearly 4,000 workers from the Philippines at various bases in Iraq in May 2004.[46] When KBR approached PPI the previous summer, the latter's CEO, Neil Helliwell, immediately reached out to Anglo-European Services (AES), the oldest licensed recruiting agency in the Philippines. According to AES's director, Gilbert "Nicky" Arcilla, his relationship with the British-born Helliwell goes back to the 1970s "when I was selling workers for him in Saudi Arabia."[47] Thus it was natural that Helliwell turned to AES when faced with the problem of assembling thousands of workers for PPI's contracts in Iraq in the span of a few months.

Scholars have generally attributed the introduction of the Philippine labor export policies in the 1970s to increased globalization and the rise of neoliberal policy prescriptions advocating deregulation, free trade, and privatization.[48] Representative of this viewpoint is Rodriguez's claim that "in a neocolonial, neoliberal state like the Philippines, labor brokerage functions to address the failures of so-called 'development.' It is a peculiar kind of 'trickle up' development as individual migrants' earnings abroad become a source of foreign capital for the Philippine state. The Philippine state remains committed to drawing direct investments from foreign capital through neoliberal economic reforms; however, it also heavily draws on 'investments' from its very own citizens."[49] While the rise of both economic globalization and neoliberal nostrums in recent decades are certainly relevant, I believe greater attention should be given to earlier policies and practices devised to facilitate the export of labor—especially military labor during the Cold War—as these provided critical institutional antecedents for later developments.

This is not to say that earlier labor migration histories have been given short shrift. As Rodriguez observes, "The labor brokerage system in the Philippines is in large part a result of the U.S. colonial legacy in the Philippines."[50] The clearest example from the colonial period concerns the sugar plantations in Hawaii. As noted above, several elements of this labor system, such as the creation of a Bureau of Labor to regulate recruiting and contracts, foreshadowed those instituted in 1974. Another pertinent case involved training and placement of nurses in the U.S., particularly through the Exchange Visitor Program (EVP), which was established in 1948. The EVP was originally intended to be an exchange program that would enable participants from around the world to work and study at sponsoring U.S. institutions before returning home. In the early years most exchanges took place with Northern European countries. But by the late 1960s Filipino nurses "began to dominate participation in the program."[51] In 1965 Congress passed a new Immigration Act, which loosened restrictions on immigration from the Philippines and other countries in Asia and established a preference system for "members of the professions and scientists and artists of exceptional ability" and "skilled and unskilled workers in occupations for which labor is in short supply." Five years later amendments to the EVP made it easier for participants to change their visa status from visitor to permanent resident. Between 1966 and 1978 approximately 7,500 EVP participants adjusted their status. Thus by the mid-1970s export of Filipino nurse labor to the U.S. was a well-established phenomenon, the significance of which was not lost on Marcos, who in a public speech delivered shortly before his 1974 presidential decree described it as "a market that we should take advantage of."[52]

Enrollment of Filipino labor by the military during the early years of the Cold War, whether as Navy stewards or logistics workers at bases across the Pacific, is also an important part of this story, and deserves greater consideration, for two reasons. First, between independence and 1974 this constituted a substantially larger flow of overseas workers than other well-known examples, such as the export of Filipino nurses. In Guam alone nearly 28,000 Filipinos were recruited to work on military projects by the late 1940s.[53] Second, as Colleen Woods and Alfred Flores have recently shown, the Philippine state worked closely with the U.S. government and private contractors to manage labor flows, developing policies and practices that were far more enmeshed with subsequent labor export institutions and law in the Philippines than is currently recognized. In 1947, for example, military construction contractor Morrison-Knudson secured permission from the Philippine Department of Labor to hire 6,000 Filipino laborers to work in Okinawa. To facilitate this process it reached out to the U.S. embassy, which negotiated an exchange of notes with Philippine secretary of foreign affairs, Bernabe Africa, that allowed military contractors to process and transport workers to "desired areas without further contact with the Philippine authorities."[54] In the case of Guam, labor recruitment was initially handled by Luzon Stevedoring, a transportation company founded by U.S. veterans of the 1898 war. Its owner, Charles Parsons, had lived in the Philippines for more than two decades and was close friends with the country's president, Manuel Roxas. According to the U.S. embassy in Manila, Parsons secured Roxas's approval to export Filipino labor to Guam in part "because of its salutatory effects on employment and the balance of payments," illustrating that political elites in the Philippines appreciated the connection between labor export and foreign currency earnings well before the advent of neoliberal globalization.[55]

The tripartite relationship between the Philippines, the U.S. military, and contractors—which Woods persuasively argues is best understood as case of "transnational imperial collaboration"—deepened under the Marcos regime.[56] In 1966 Marcos asked about the possibility of Filipino firms obtaining special consideration for military contracts in Vietnam.[57] While this was rebuffed by the U.S. government, construction contractors like RMK-BRJ did turn to the Philippines as one of their main sources of foreign labor. By 1969 approximately 20,000 TCNs (primarily Filipinos and South Koreans) were working at military bases in Vietnam.[58] Almost 1,000 more were employed at Poro Point in the Philippines, constructing portable piers that were then towed to Vietnam for use as temporary port facilities.[59] In contrast to Guam, where the principal attraction of Filipino labor for contractors was that they served as a low-wage workforce paid less than half the wage of native Chamorro employees, the TCN workforce in Vietnam consisted in the main of well-paid skilled laborers, earning—according to a survey conducted by the U.S. embassy—about $6,700 a year.[60]

In 1968 the U.S. and Philippines concluded an offshore labor agreement that provided guidelines on recruiting and employing Filipino citizens by the U.S. military and its contractors "in certain areas of the Pacific and Southeast Asia."[61] The agreement includes provisions concerning contractor and recruiting documentation, remittances, transportation procedures, and employee benefits, including a minimum wage, holiday pay, vacation and sick leave, health insurance, severance pay, living quarters arrangements, and Philippine Social Security benefits. Although the agreement still remains in force, a 1992 ruling by the Fifth Circuit Court of Appeals in the U.S. effectively gutted most of its provisions by deciding that the enumerated employee benefits are required for Filipinos directly employed by the U.S. military, but not its contractors. The decision rests on a distinction in the agreement between "employer" (understood as "United States military forces") and "contractor" (defined as "enterprises . . . under contract with the United States military forces . . . who may wish to recruit Philippine citizens in the Philippines for employment or re-employment in the offshore areas defined herein"). Though the agreement states that "employment contracts between contractors and Philippine citizens shall be consistent with the standards and terms established in this Agreement," the Fifth Circuit found that responsibility for ensuring this consistency rests with the Philippine government, not U.S. military authorities.[62] Despite this later court decision the 1968 offshore labor agreement remains a direct—if little-known—precursor to the labor export system established by Marcos's 1974 labor code, and the first of many bilateral agreements signed between the Philippines and labor-importing countries.[63]

Profiting from War

Just as Marcos sought favorable conditions for Filipino firms and labor in the Vietnam War, from the very outset of the U.S. invasion of Iraq Arroyo positioned her foreign policy to curry favor with the U.S. in hopes that Filipino companies would profit from the anticipated postwar reconstruction bonanza. Indeed, despite a distance of four decades, the parallels between these two episodes are striking, demonstrating both the durability of imperial and client-state formations and that "the foreign policy of the Philippines is intimately connected with overseas employment."[64] After Marcos was elected president in 1965 he decided—against substantial domestic opposition—to send a battalion of noncombat (engineering) troops to Vietnam as a show of support for U.S. war efforts. In exchange the U.S. agreed to increase economic and military aid to the Philippines, while military contractors set up shop in Manila, recruiting Filipino labor.[65]

Likewise, Arroyo joined the "coalition of the willing" in Iraq despite widespread domestic opposition and also agreed to symbolically support U.S. war efforts by sending a small contingent of noncombat troops. And as did Marcos decades before, she justified this decision in economic terms, arguing in April 2003 that companies from countries that joined the coalition would "get first crack at the development efforts." Even if this was not the case, her labor and employment secretary added, "I'm confident that if they're looking for skilled workers, they'll come to us."[66] A few days after Arroyo assured reporters that the Philippines would get "first crack" at postwar work in Iraq, she issued an executive order establishing a Public-Private Sector Task Force on the Reconstruction and Development of Iraq. This task force was charged with facilitating "the participation of Philippine companies in the rehabilitation and development of Iraqi infrastructure" and developing "procedures to expedite deployment of Philippine manpower and other services in the fulfillment of contracts."[67] Three weeks after the order the task force's head, Roberto Romelo, was openly bullish on the potential windfall that Iraq represented. Traveling to Kuwait to pitch Filipino labor to U.S. contractors, he suggested that 30,000 workers in Iraq would be a reasonable baseline, due in part to the well-established Filipino labor presence in the region. "I'm quite optimistic because we have a track record. . . . Every one of the prime contractors I've spoken to say, 'we've dealt with you before in the Middle East and we look forward to working with you again.'"[68]

Another factor in their favor, Romelo argued, was the rapid deployment of Filipinos the previous year to Guantanamo. "They needed it right away. Within one week we had people on a plane and on the way to Guantanamo."[69] The "they" he was referring to was the U.S. government, which reportedly reached out "directly to the office of President Gloria Arroyo" in March 2002 asking for help in quietly assembling a team of 400 engineers and construction workers that were tasked with building Camp Delta, the main detainment facility for extrajudicial prisoners in the "war on terror."[70] As it would the following year, the Philippine government worked with AES to speed up the hiring process for the benefit of PPI, KBR's subcontractor for the Navy construction contract. Eventually the Filipino workforce on Guantanamo would grow to approximately 1,500 in number.[71]

Recruiting agencies like AES were among the biggest supporters of Arroyo's decision to participate in the postwar occupation of Iraq. They were also the fiercest critics of her imposition of the travel ban in 2004, arguing that it violated the rights of those who wished to work in Iraq despite the risks.[72] The agencies, led by AES, organized a series of public protests against the ban in August and September in a futile attempt to force the government to reverse course. Despite this both AES and PPI were feted by the Arroyo administration the following year for their success in facilitating the export of Filipino labor to Iraq, with AES given a "Top Performer" award that recognizes especially productive recruiting agencies and PPI presented the "International Employer Award" for "displaying continuous preferences for Filipino workers and providing them with excellent career advancement and a generous package of employment benefits."[73]

Ten years after this ceremony, while I was finishing research in the Philippines, interviewees were abuzz with news that recruiting agencies in Manila would soon be looking to source up to 400 workers to support U.S. military operations at Al Udeid Air Base in Qatar. The headquarters for military operations in CENTCOM, Al Udeid was also the epicenter of the rapidly expanding air campaign against ISIS in Iraq and Syria.

THE WAGES OF PEACE AND WAR

To tell you the truth, Bosnia should be grateful for George Bush. Because of the wars in Iraq and Afghanistan there's around 10,000 people working there.

—Fedja

It is a beautiful July afternoon in northeast Bosnia. I am sitting in a fashionable café in Tuzla preparing for an interview with a former KBR employee named Goran. After a few minutes he arrives, wearing an expensive Oakley watch and a big smile. We order drinks and he starts telling me his story. Goran worked for KBR in Iraq for four years, beginning in 2006. Before this he was in law enforcement, but "it was barely a survivable salary. Like six hundred [Bosnian] marks a month. . . . I was 28 years old and I didn't have my own car, I didn't have my own apartment."[1] Like so many others he saw little chance of his situation improving if he stayed in Bosnia. When he heard that KBR was in Sarajevo recruiting, he leapt at the chance. "When they came and [were] like, 'You want to make $56,000 or $60,000 dollars a year?' Fuck yeah, man! And if they offer me anything . . . I mean, I didn't go there to pick a job. I'm going to take whatever they find suitable. So I went there as a labor foreman." The reason for Goran's excitement is not difficult to understand. Working for KBR in Iraq offered the opportunity to make roughly twenty times his Bosnian salary.

In a country with a GDP per capita of less than $5,000 and an official unemployment rate that hovered around 30 percent in 2006, Goran was not alone. Since 2002 thousands of Bosnians have worked for military contractors in the Middle East, Afghanistan, and Africa. Thousands more also worked in support of U.S. peacekeeping forces in the Tuzla region in the latter half of the 1990s, many employed by Brown & Root as cooks, cleaners, drivers, administrative staff, and construction laborers. Indeed, it is no exaggeration to say that for the past two decades the U.S. military has—directly and indirectly—been the most significant source of employment opportunities for people in Bosnia. This reflects both the country's moribund postwar economy and the scale of logistics contracting by the military, beginning with its peacekeeping missions in the Balkans.

Military labor undoubtedly pays well compared to most work in Bosnia. But the experience of those who have chosen this path is also marked by precarity, both in relation to the work itself and the marginalization of one's social and economic position in Bosnia. This is especially the case after the contract ends and people return home. Goran, for instance, was able to buy an apartment and car, help pay for his sister's university expenses in the U.S., and support his parents, who live on a small pension, but he was unable to return to his previous position with the police. After months of searching he found a job with a local travel agency, working for commissions. However, few in Tuzla can afford expensive vacations, especially now that the money flowing in from the Middle East and Afghanistan has slowed to a trickle. Gradually all of his remaining savings melted away. "I remember the exact day when I hit the last hundred dollars on my account," he recalls. "I already started working for the [travel] agency, but didn't have any work [commissions]. Even now I'm not making any money. It's just making ends meet. That's it. When I saw my last hundred dollars in the account I had an anxiety attack . . . How am I going to live? Yeah, that's when you think about going back." As I describe below, the struggle to readjust to life in Bosnia, coupled with thoughts of finding another job abroad, is common among Bosnians who have done this work.

This chapter traces the impact of military contracting on the social and economic fortunes of individuals and communities in Bosnia over the past twenty years. I begin by outlining the employment of Bosnians as LN labor for U.S. peacekeeping forces in the late 1990s and early 2000s, explaining why this was concentrated in the northeast of the country. I also argue that this phenomenon needs to be situated within an analysis of the broader peacekeeping economy of postwar Bosnia. Following this I describe the shift to working for contractors in Iraq and Afghanistan, a shift that often led to a dramatic upgrade in pay and status—at least initially. Finally, I explore the duality of prosperity and precarity experienced by Bosnian workers, with a focus on how the latter has been profoundly shaped by social, economic, and political conditions in Bosnia.

The Peacekeeping Economy in Postwar Bosnia

In December 1995 U.S. Army soldiers serving as part of the multinational Implementation Force peacekeeping contingent crossed the Sava River into Bosnia.

They were accompanied by Brown & Root which, as noted in chapter 2, was the company contracted under the LOGCAP program to provide logistics support. The primary motivation for using LOGCAP, according to one of the chief planning officers, was the force reduction caps imposed on the military "JCS [Joint Chiefs of Staff] and the President defined some force caps on the number of troops we could have. That posed our next dilemma because our troops-to-task estimate was well above the 25,000 troop cap for total U.S. commitment. . . . LOGCAP was immediately identified as one of the methods by means of which we could reduce the dependence on uniformed service members and meet our construction and service requirements within the force caps we were being asked to accept."[2] Brown & Root was charged with providing a broad range of support, including base camp construction and maintenance, laundry, showers and latrines, food service, bulk fuel storage, and transportation of supplies into the country and among dozens of camps.

To carry out these tasks Brown & Root relied heavily on local labor. Almost immediately upon arrival in Bosnia it started recruiting as it scrambled to set up and manage three initial "force provider" camps in Tuzla and the nearby town of Lukavac, which would become the primary logistics depot in the region. Thus from the beginning one's chance of getting this work was shaped by the contingent fact of where you lived. Sanja, a college student from Lukavac with little more than knowledge of English at the time, illustrates this dynamic:

> Everybody has a different story. My story wasn't nice. My mother got sick. I was in the first year of college and she got sick, breast cancer. It was '95. It was almost the end of the war. The situation was all bad. No money, nothing. You know how it was. Actually you don't know. You weren't in the war. But anyway, we needed money and I had to find a solution. KBR was here [Lukavac]. The military was here. So I applied— actually it was not KBR it was Brown & Root at that time—and I got a job in the coke plant [in Lukavac]. So they had a camp in there and I started to work for the MWR [Morale, Welfare and Recreation center]. They had a library in the MWR and that's where I started.

Another successful applicant, Djenan, met two Brown & Root human resources employees in the lounge of the Hotel Tuzla, where his uncle worked. They told him to come to their office in Lukavac if he was interested in a job. Speaking with a southern Texas twang and colloquialisms picked up during nearly fifteen years' work with U.S. contractors, he describes the chaotic scene at the office in early January 1996:

> It was a small office where all the small shops [in Lukavac] are. I was like, "This is the company?" I didn't really realize the magnitude of it

yet. There were a shitload of people out front. Everyone in Lukavac knew [they were hiring] by then. Somehow I managed to get through the people and knock on the door. I was waving some resumes of mine. They let me in and I went upstairs and had an interview with these two guys. It went on for like 20 minutes. They asked me all sorts of questions. My English wasn't near as good as it is today, but it was ok, and they said "Ok, we'll give you a call." I was walking away and thinking, "What is going on?" You see they had all these [Bosnian] rednecks coming in too. I didn't realize they were plumbers and what have you. . . . Sure enough they called me a few days later. It was like, "Be in Lukavac at 7 a.m. [tomorrow] at the cultural center."

Within months Brown & Root, Navy Seabees, and Air Force Red Horse engineers built a network of more than two dozen camps across northeast Bosnia (figure 4.1). The center of this network was a series of bases scattered around the Tuzla region, anchored by a former Yugoslav air base southeast of the city. In a short period of time this would be transformed into Eagle Base, the main base of operations for U.S. forces in Bosnia. A second cluster of camps were established north of Tuzla in the Posavina region to enforce the demobilization of the armed forces of the Federation and Republika Srpska (RS) along the Inter-Entity Boundary Line (IEBL) that divided Bosnia's two substate political entities. The Posavina region contained some of the most bitterly contested territory of the war, especially the corridor running through Brčko that connected the two halves of the RS, which was considered one of the most likely potential flashpoints for renewed conflict due to its strategic importance and still-unresolved status after the war.[3] A third, southern band of camps stretched along a strategic road connecting Vlasenica in the RS and Kladanj in the Federation, and the road extending south from Kladanj to Sarajevo. Finally, U.S. forces and Brown & Root contractors set up logistics hubs in Hungary and Croatia, and were based at Butmir, the multinational peacekeeping headquarters in Sarajevo.

The location of U.S. bases in Bosnia was a product of the decision to divide peacekeeping responsibilities in the country into three zones. In addition to the American zone in the northeast, British troops led operations in northwest Bosnia, and French troops were based in the south. Nordic, Russian, and Turkish peacekeepers also manned sites to the west and east of Tuzla in the American zone. After the first year of peacekeeping operations most of the small outlying camps were closed, leaving U.S. forces even more concentrated around Tuzla.

As noted in the introduction, the peacekeeping missions in the Balkans in the 1990s (Bosnia and Kosovo) were the first time when the number of contractors are estimated to have reached parity with deployed troops. The majority of these were local hires. According to one study commissioned by the military, Bosnians made up 80 percent of Brown & Root's workforce in the country.[4] In addition to this they worked in base post exchange (PX) offices and as interpreters for the Army, with 300 employed in the latter role by the end of 1996 according to former interpreters I have interviewed. Brown & Root and the military also contracted with local firms for supplies of construction materials like gravel and lumber, transportation equipment, and the completion of various infrastructure and reconstruction projects. In April 1996, for instance, the military's regional contracting office in Tuzla signed a contract worth 2 million deutsche marks with the firm Tuzla Putevi for repair work on the road between Tuzla and the Croatian border.[5] In short, though there is no hard data on the total number of Bosnians in the Tuzla region working for the U.S. military and the various contracting firms supporting it in the years immediately following the war, I believe a conservative estimate would be more than 10,000.[6]

It is useful to view this workforce through the lens of what Kathleen Jennings calls the "peacekeeping economy," which she defines as "the economic multiplier effect of peacekeeping operations via direct or indirect resource flows into the local economy."[7] According to Jennings, there are several elements that constitute the peacekeeping economy. The first is formal employment with international organizations and peacekeeping forces. Following the war, thousands of Bosnians were hired as project officers, interpreters, or support staff by major international organizations operating in the country such as the Organization for Security and Cooperation in Europe (OSCE), the Office of the High Representative, and the UN. Like work for military contractors, this constituted a significant and distinct employment sector in the country.[8] In addition to formal employment, the peacekeeping economy also consists of informal work for international staff; the development of industries that cater to internationals like restaurants and bars, hotels and apartments, and the sex industry; and investments in postwar reconstruction of infrastructure and housing.[9]

The peacekeeping economy constitutes a significant portion of economic activity in the immediate years following a war, especially in small countries like Bosnia that host a sizable international presence. At the same time its effects are also highly uneven spatially, as they tend to be concentrated in the national capital and cities where international organizations and peacekeeping troops are located. Consider the impact of the peacekeeping economy on Tuzla in the late 1990s. The first thing to note is that the city and its surrounding region is relatively small, with less than 500,000 people living in Tuzla Canton. Tuzla itself is one of the oldest inhabited settlements in Bosnia, due to its saline lakes that have been utilized for salt production for centuries. During the time of socialist Yugoslavia it became an industrial city known for coal mining, chemical production,

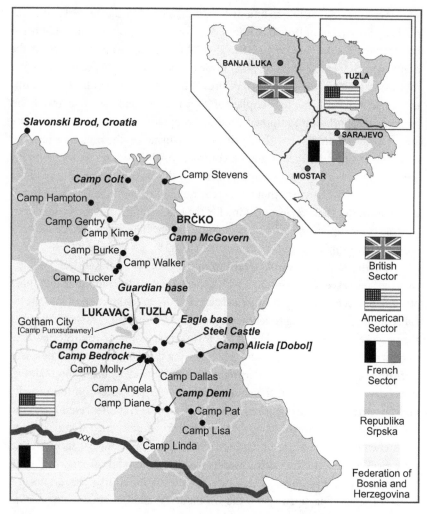

FIGURE 4.1. Multinational peacekeeping sectors and main U.S. peacekeeping bases in northeast Bosnia, 1996 (italicized and bolded sites used beyond 1996)

and metal working. These jobs offered security in employment and housing. Following the war, many of these industries struggled to regain their footing due to a combination of wartime destruction and theft of infrastructure and equipment, disinvestment, and lack of competitiveness on international markets. As a result, industrial centers like Tuzla faced especially difficult economic conditions. Thus it is difficult to overstate the impact that the rapid recruitment of roughly 10,000 people working in support of U.S. forces, as well as thousands more employed by international organizations, and the burgeoning service economies offering food, entertainment and housing for international staff—including the emergence of a large sex industry catering to international civilians and peacekeepers—had on the local economy.[10]

Going to War

Significantly for those in the Tuzla region, just as the peacebuilding intervention in Bosnia was beginning to wind down in the early 2000s U.S. military activities in the Middle East were ramping up. So the transition from work in Bosnia's peacekeeping economy to a distant war economy was, for many individuals, made without a significant break in employment. One-third of those I interviewed in Bosnia, for instance, started out as Brown & Root employees in the late 1990s. Several more served as military interpreters or worked for organizations like the International Committee of the Red Cross and OSCE before finding positions with military contractors in the Middle East and Afghanistan.

Brown & Root employees who followed the company (then known as KBR) to Iraq and Afghanistan in the early 2000s experienced a distinct upshift in pay and status. In Bosnia they had been classified as LNs and paid wages that were linked to local salaries. According to those I interviewed, depending on skills and job category wages with Brown & Root in Bosnia ranged from one and a half to three times the average salary in Bosnia. This pay range roughly corresponds with Catherine Baker's research on local military interpreters with the British Army in Bosnia, who earned two to three times the going wages.[11] Salaries paid by KBR in Iraq and Afghanistan were much higher than this, where even individuals with no previous experience, like Goran, were able to earn more than $50,000. Long-time employees recruited for positions requiring technical skills and/or experience, like quality assurance/quality control (QA/QC) supervisor or procurement officer, could make up to $80,000.

The increase in pay was also accompanied by an increase in status. As LNs, Bosnians working for U.S. peacekeeping forces were not able to shop at the PX stores or use MWR facilities during off hours.[12] They were also subject to repeated security screening procedures by the military, with one worker recalling that these occurred "every six months, for three to four hours." And when entering and walking around bases their movement was closely monitored and circumscribed. To give one example, in May 1997 an article titled "Be Not Afraid" appeared in *The Talon*, a weekly newspaper produced by the Army's public affairs office at Eagle Base, describing a demonstration of military police dog capabilities. The purpose of this demonstration was to reassure local Brown & Root employees who found the dogs "frightening" and "aggressive" during searches and patrols on bases.[13]

In Iraq and Afghanistan the military classified Bosnians as TCNs, but as KBR employees they stood apart from and above others working for the company's subcontractors. When I asked Elvis, a former military interpreter in Bosnia who began working for KBR in the 2000s, whether the label TCN was ever applied to KBR's Bosnian workforce in Afghanistan, he replied, "In KBR I heard the term OCN [Other Country National] or TCN, which is the same shit, maybe like three times in four and a half years. . . . You were a KBR employee and you were treated as such, unless your point of origin was a matter of statistics." As a KBR employee Elvis enjoyed a variety of privileges, including possession of a Common Access Card (CAC), issued by the DoD or badges that allowed access to military dining facilities (DFACs), MWR centers, and PX stores.

The 2008 decision to split the LOGCAP IV contract among three firms—KBR, Fluor, and DynCorp—stimulated a second wave of hiring in the Tuzla area. Under this new contract KBR retained logistics support in Iraq and the Gulf states but Fluor and DynCorp were given responsibility for operations in Afghanistan. Both companies faced a need for labor, especially following President Obama's decision in favor of a troop surge in Afghanistan in 2009. Fluor and DynCorp quickly set up recruiting offices in Tuzla. The response was remarkable, as illustrated by the following vignette from Larisa Jasarevic, an anthropologist who studies debt, divination, and informal markets in postwar Bosnia, about a 2011 visit to a well-known fortune-teller in Tuzla:

> I have been casually visiting Zlata since 2006. In 2011, I found the cups [used for divination, reading Turkish coffee remains or 'mud'] much larger and the scope of her vision extended to keep up with the migration of economic opportunities, from regional, largely informal market trade to more transnational pursuits of fortune with American defense contractors (KBR and Fluor International) in Afghanistan and Iraq. Among those who seek her out . . . many work for or are applying to Fluor International or else dating, desiring, marrying, and otherwise caring for men employed or seeking employment in Afghanistan and Iraq. A young woman, anxious about her protracted engagement, walked out of Zlata's room with assurances about the date for her wedding and, just as exciting, news of her future husband's job offer in Afghanistan.[14]

One Bosnian magazine described the phenomenon another way in 2009 when it claimed that the "mass departures to Afghanistan" of people from Lukavac represented their "answer to the recession" in Bosnia.[15]

At the same time, jobs with Fluor and DynCorp were usually accompanied by a reduction in pay and status relative to U.S. contractors and other TCNs. The reason is that for the new contract (LOGCAP IV) the Pentagon directed its prime contractors to bring salaries for direct hires more in line with prevailing wages in countries that they come from—or at least a more reasonable premium to prevailing wages than what had been paid by KBR under LOGCAP III. DynCorp, for instance, classified its employees according to four categories: 1) Expats (Americans), 2) Foreign National United Kingdom (FNUK), 3) Foreign National European (FNE), and 4) Foreign National Asian (FNA).[16] Pay and privileges were roughly equivalent for expats and FNUK employees. My interviews suggest that DynCorp paid FNE workers (which were primarily from Bosnia, Macedonia, and Kosovo) less than KBR, with salaries for most positions between $30,000 and $50,000. This, of course, was still far more than one could earn in Bosnia—if you could find a job. Fluor also set up a tiered classification system that distinguished between company staff, Americans hired on contract, West European employees, East Europeans, and workers from Asia. Under this system Bosnians and other workers from the Balkans (tier IV) earned 45 percent of what Americans and West Europeans were paid for the same jobs.[17]

Fluor's and DynCorp's classification schemes did not just govern pay, they also reflected company cultures that set their American and West European workers apart from Balkan and Asian employees. According to Elvis, who also worked for Fluor in Afghanistan, "Fluor was very insensitive toward that. It was OCN this, OCN that." Damir, who also worked for both KBR and Fluor, highlighted the difference between the two companies with the following story:

> DAMIR: I think in KBR all guys were the same, you know. Americans, Bosnians, Macedonians. Some guys from Europe. All were the same.
> ME: Do you mean same in terms of money or same—
> DAMIR: The rules were all the same. Rules. For Americans, for me. In Fluor it was not the same.
> ME: Can you explain?
> DAMIR: The first time I went with Fluor and landing in Bagram, some HR [Human Resources] guy from Fluor come pick us up. And that flight had some American guys, some Bosnians, Macedonians, some Filipinos, some countries from Asia. And that American guy [in HR] came and said, "American guys go on that side. And some Europe guys—from Germany or France—go with American guys." And then he said "Bosnians, Macedonians, and Asians, go on the other side." And first bus came to pick up American guys and second bus came to pick us up. And at that time I see it is not the same as KBR.

Disgusted with Fluor's treatment of longtime former KBR employees from Bosnia, Damir worked just two months with the company before returning home.

The greatest decrease in pay and status was experienced by Bosnians who were recruited on "Asian" contracts with DynCorp. This practice began in 2010 according to several former DynCorp employees I talked with. The typical Asian contract paid between $900 and $1500 a month—less than many who worked for KBR as local employees in Bosnia in the 1990s made. Recruiters also falsely promised that Bosnians would be able to switch over to a "European" contract when they arrived in Afghanistan. One applicant, Diana, was told, "After three months, you can change the position. You're not going to stay on this position. I can guarantee you that." It took her two years to obtain a European contract. Another Dyncorp employee, Edin, recalls: "When I finished one Asian contract, I asked them, 'Are you going to give me now European contract, because I'm from Europe? Maybe Bosnia is not in European Union, but it's still in Europe.' They said, 'No way. If you're going to sign this one, sign. . . . If you don't want to sign, we're going to buy you a ticket [home].' I said, 'Ok, buy me the ticket. Put me on the first plane. I want to go home.' They said, 'It's no problem. Just go in your tent. They're going to call you tomorrow and give you your ticket.'" Despite the dramatic reduction in salary, DynCorp did not find a shortage of applicants willing to work in Afghanistan on Asian contracts. In 2010 a Bosnian magazine estimated that more than 5,000 people from the Tuzla region were working in Afghanistan and Iraq, with thousands more looking for the chance to join them.[18] For many, clearly, the opportunity to work abroad—even in a warzone for as little as $900 a month—was preferable to struggling to survive in Bosnia's depressed postwar economy.

The Duality of Prosperity and Precarity

The intersection of well-paying but precarious work and the more general condition of precarity in postwar Bosnia has produced a paradoxical duality of prosperity and precarity for Bosnians involved with military contracting over the past twenty years. One sign of prosperity is the construction of several new apartment buildings in Lukavac and Tuzla that are informally called "Iraq" and "Afghanistan" due to the large number of people who have worked in those countries who have purchased flats, often with cash. Another is the consumption of luxury goods like the watch that Goran wore to his interview. Expensive watches seem to be especially popular status symbols with Bosnian men. One individual I talked with spent three months' salary on a Tag Heuer watch in

Dubai during his first leave from Afghanistan. His wife, he recalls, was less than pleased. Vacations to the Adriatic coast are also popular among workers, who receive a month's leave from KBR and Fluor three times a year.[19]

The conspicuous consumption of goods like cars, vacations, expensive watches, and clothes is frequently remarked upon. Indeed, one does not need to spend much time in the Tuzla region to pick up an undercurrent of resentment toward those who have worked in Iraq and Afghanistan, mixed with criticism that many have squandered their money on frivolous purchases. Yet most people I talked to spent the bulk of their earnings on more prosaic things such as housing, helping their children attend university, supporting parents who live on meager pensions, or giving money to siblings and extended family members who are struggling to get by.

For those like Sanja and Djenan who joined Brown & Root shortly after the Yugoslav wars and then followed KBR or other firms to the Middle East and Afghanistan in the 2000s, this work essentially constitutes a professional career, spanning the majority of their working lives. After finishing his contract with Fluor for the Ebola mission in West Africa in 2015, Elvis, for instance, had spent nearly two decades working for the military or one of its logistics contractors. In addition to amassing career earnings far larger than possible for all but the most fortunate—or politically connected—in Bosnia, nearly all of the longtime KBR employees that I spoke with also appreciated the chance to earn promotions and pay increases, as the following exchange with Esad illustrates:

> ESAD: The company [KBR] was great to us. It is very hard to find a company like that here.
> ME: In what ways was it a good company to work for?
> ESAD: Giving us an opportunity to prove ourselves. To—how to say— you had a lot of opportunities working with that company. Where you started and where you finished. There was no discrimination. If you are smart and can do your job you can move up.
> ME: So how many times did you apply?
> ESAD: How many times did I get promotions?
> ME: Yeah.
> ESAD: Hmmm . . . five or six. And I applied for maybe ten or fifteen jobs, I can't remember exactly.

Esad began working with Brown & Root as an "assistant truck driver with SST, trash." Basically his job was to hook up the "shit-sucking trucks" to latrine tanks and make sure they did not hit equipment or buildings when backing up and navigating the main camp in Lukavac. By the end of his time a decade later he was working in KBR's administrative center in Kuwait "putting reports together" on the company's operational activities in Iraq and Kuwait. This type of career trajectory is not uncommon among Bosnians who began working with military contractors in the 1990s.

On the other hand, there are several aspects of the peacekeeping and war economies that fuel precarity, both in relation to the work itself and the marginalization of one's social and economic position in Bosnia. The first, obviously, involves the risk of severe injury or death while working in an active warzone, as evidenced by the death of multiple individuals from the region. Another factor is the highly contingent nature of employment—both in regard to the ubiquity of short-term contracts and the fact that workers can be immediately terminated for violating any one of myriad rules regulating life on military bases. This leads to a situation that Catherine Baker, who has researched the position of military interpreters in postwar Bosnia, aptly calls "prosperity without security."[20] Additionally, salaries paid by both peacebuilding organizations in Bosnia and military contracting firms abroad do not include contributions to state employment or pension funds. This means that workers are not able to build up credits for retirement benefits. Nor are they eligible for unemployment benefits when their job ends.

As important as these factors, though, are other more existential dimensions of precarity linked to the peacekeeping and war economies, especially the ways in which this type of work socially and economically marginalizes individuals in Bosnia, which presents a number of challenges when their contracts end.[21] One way this occurs is through a social distance developed through enculturation of the mind-sets and business practices of foreign colleagues, organizations, and companies. Longtime workers like Djenan do not just talk like Americans, they have also picked up different habits and ways of thinking after interacting with U.S. troops and civilians for years on end that make it difficult to reacclimate themselves to life in Bosnia. Tatijana, who has spent her entire life working for international organizations in Bosnia and KBR in Iraq, highlighted this as her biggest challenge.

> I'm not even sure if I could function in a work system around here. I've honestly never worked for a local company in my entire life. I'm not just talking about the money. It is just the way things work. The efficiency of it. Around here it's like, yeah, we'll get to it. You'll get your money when you get it. I'm used to working a system where I know I have to do this, this, this, and this and do it well to be able to keep my job and get my money at the end of the month. Well, the economy around here sure

doesn't function like that. It's who you know, who you're related to, and stuff like that. That's how you get a job and that's how you keep it. It doesn't really matter what your qualities are or what you bring to the table with your experience and skills and knowledge. It's about networking and, basically, it's obvious. It's nepotism.

Echoing her, Sead, who worked for DynCorp in Afghanistan, insisted that his relatively brief time with the company was a blessing: "I didn't stay too long over there. It was only two years and two months. But I know people who stayed more than five years. And when you come home you're just lost. You don't have too much contact with your other friends. You don't have contact with your mom, or maybe with the wife or with your child. Because over there you change as a man, a person, like another person. I think it is bad if you stay too long over there." Samir, who has worked for several international organizations in Bosnia over the past two decades, put this issue to me most succinctly and poetically when he stated, "After fifteen years with the IC [international community] we don't belong here [Bosnia] any more. We are an in-between people."

Another aspect of marginalization people report is that their experience and skills are not valued by Bosnian employers—who also fear that they won't work for low-wage salaries—making it even harder to return to the local economy. Ivan, who worked for KBR and DynCorp for sixteen years starting in 1996, told me: "The problem is . . . nobody's going to employ a man who is 40 years old without any kind of experience. Local experience. There is a kind of, how do you say, I can't find the word, the locals they do not like people who worked for rich companies. They think, you're full of money, you don't need a job, 'Why [do] you need a job? You just spent 10 years working for KBR earning $7,000 to $10,000 bucks per month.'" After struggling to make a living in Lukavac, he returned to Iraq to work for the logistics services company, Sallyport, in 2016. This perception was also articulated to me by a business owner in Tuzla, who bluntly explained why he tends not to hire those who have worked for PMCs in the Middle East: "You have somebody that spent ten-plus years abroad. He lost the feeling of things in Bosnia. Completely useless. He got used to being paid quite a lot. He cannot get paid a lot here. So the motivation for the job is questionable. . . . [He] is probably waiting for another project to go off [to]." Esad's experience after returning home in 2007 illustrates the struggle that many returning workers say they face:

> ME: What was the biggest challenge [when you came back home]?
> ESAD: For two years I was applying for jobs in Bosnia. And I didn't even get an interview. That was the biggest challenge.

ME: What kind of jobs were they?

ESAD: They were logistics, transportation. Even as a truck driver. And I did not get interviewed.

ME: Did people tell you why?

ESAD: Yes, they tell you. They say because of incomplete paperwork, or say, "You don't have the experience."

ME: They don't count experience with KBR?

ESAD: No, they don't . . . You have to have about a year or two years' experience in the field you are looking for. But in Bosnia. And I don't have it. I spent twelve years with KBR.

ME: So they don't count your work here in Bosnia with KBR [as experience] either?

ESAD: No, they don't. I don't know why.

After years of applying he eventually found work as a cab driver. Nearly everyone he knows is in a similar situation. "Sometimes at coffee when I meet people who were over there, we ask, 'Have you found a job?' And everybody is depressed because they haven't found a job. It is miserable."

One alternative to working for someone else is to open your own business. But as Enis, who worked for Fluor in Afghanistan for several years, explained to me, in a country like Bosnia this too has its downsides.

I was like, "Ok I am going home and I got some money saved. And I am going to open my own business and live off it, and that's it." But [there are] so many risks to opening your own business. From the state—papers, laws, unethical competition. [And then] criminals and security. So it really, the time is so bad that I do not dare to invest in anything. Because if I slip then I am fucked. The other day a friend who runs his own business—printing, making advertisements—said, "Enis, listen to me. I am your friend. If you want to open anything, open a cold beer and shut the fuck up and enjoy it." Because when he showed me his business, how much people owe him, or how much he owes to his suppliers, it's a vicious cycle. It's hard for him just to somehow stretch enough to pay the guys that work for him, or pay the taxes to the state. So I am looking around, applying to local companies.

The challenge of reintegrating with society extends beyond work, as illustrated by the following quote from Srdjan, who has worked for several military contracting companies in Bosnia, Iraq, and Afghanistan since 1995: "The main struggle is to get resocialized back into civilian life. Especially in Bosnia: unemployment, the political situation, missed growing of children, failed marriages. *Facing the reality of this life here.* And some guys just can't find themselves. And again they apply for another mission. . . . Because unfortunately there aren't many options here, to get employed, have a regular life. So people, after some time they get disappointed" (italics mine). The phrase "facing the reality of this life here" is significant for a couple of reasons. The first is that I heard variations of it from multiple people when describing their current struggles after returning from Iraq or Afghanistan—struggles that echo those experienced by returning veterans in the U.S. More important, though, is how the phrase succinctly references the general experience of precarity in postwar Bosnia that is linked to a range of political, social, and economic conditions.[22] That is, precarity in reference to not just a postsocialist economy marked by high unemployment and the loss of economic security, but also to endemic ethnonationalist rhetoric, political uncertainty, and the ongoing struggle to return to a "normal life" in the aftermath of violent ethnic cleansing and displacement.[23] One example of the pervasive experience of existential precarity in Bosnia is provided by a stunning 2017 news story by the journalist Gordana Knežević that examines the increasingly widespread use of antidepressant and antianxiety drugs. According to medical statistics Knežević cites, in a country of roughly 3.5 million people there now are 4.3 million prescriptions for the antianxiety drug bromazepam and more than a million prescriptions for various antidepressant medications.[24]

In 2014 protests against political dysfunctionality, corruption, unemployment, and unpaid wages and pensions by publicly owned companies erupted in Bosnia. Nationalist political parties' offices, and government buildings—including the Presidency Building—were set on fire while tens of thousands marched in cities across the country.[25] The initial site and epicenter of protests was Tuzla. Several scholars have argued that this can be explained by the fact that Tuzla is an economically depressed former industrial city that has also been a center of left-wing, anti-nationalist politics in the country since the early 1990s.[26]

I believe that in addition to their location, the timing of the protests is also explainable, in part, by the fact that over the two years prior to 2014 employment in the distant war economy contracted in conjunction with the drawdown of troops in Afghanistan. While the loss of employment of a few thousand individuals may seem small, the multiplier effect of these jobs in Tuzla is significant. As Enis pointed out to me, "Try to imagine for a city or area like this, when you have one company delivering 5,000 paychecks every week, multiplied with their families. So like 20,000 people directly connected, or earning, putting bread on the table. And it's gone." In addition to supporting multiple family members, the earnings and consumption of workers have also boosted a variety of industries in Tuzla, from construction and real estate, to auto sales, restaurants, and travel agencies. This injection of money from the peacekeeping and war economies over the previous twenty years masked, to an extent, the degree of economic precarity in the region. As this money has dried up in recent years, frustration with "the reality of this life here" has mounted. In the end, the temporary prosperity presented by military labor has not offered an escape from political and economic precarity in Bosnia.

Part 2
ROUTES

SUPPLYING WAR

Strategy, like politics, is said to be the art of the possible; but surely what is possible is determined not merely by numerical strengths, doctrine, intelligence, arms and tactics, but in the first place, by the hardest facts of all: those concerning requirements, supplies available and expected, organization and administration, transportation and arteries of communication.

—Martin van Creveld

The U.S. military's ability to project force across the globe rests on the immense logistical resources it can bring to bear, without which the variety of operations it has carried out since the end of the Cold War would not be possible. Yet it is the rare analysis of warfare—now or in the past—that gives sufficient attention to the import of logistics. One exception is a classic, but little-known, text on the topic written a century ago by Marine Corps colonel George Thorpe, who drew upon the analogy of theater to illustrate the key role that logistics plays:

> Strategy is to war what the plot is to the play; Tactics is represented by the role of the players; Logistics furnishes the stage management, accessories, and maintenance. The audience, thrilled by the action of the play and the art of the performers, overlooks all of the cleverly hidden details of stage management. In the conditions now adhering to the drama it would hardly be incorrect to assert that the part played by the stage director, the scene shifter, the property-man, and the lighting expert equals, if it does not exceed in importance, the art of the actor. . . . Logistics is the same degree of parvenu in the science of war that stage management is in the theater.[1]

As Thorpe perceptively noted, stage management depends on a diverse pool of labor and expertise. Logistics, he argued, is also a multifaceted enterprise with activities ranging from transportation of supplies to care of wounded troops.

Conducting the wars in Iraq and Afghanistan, as well as rapidly expanding counterterrorism operations in Africa, has involved the movement of a tremendous amount of goods and people along lengthy and complex supply lines, the construction and maintenance of hundreds of bases—many the size of small cities—in remote and challenging environments, and the provision of a panoply of life support services like food, laundry, showers, and billeting for service members. Consider the remote Arba Minch drone base in Ethiopia that was operational between 2011 and 2015. According to military documents, two medium altitude drones, one MQ-9 Reaper and one MQ-1 Predator, flew from this facility, providing intelligence, surveillance, and reconnaissance (ISR) coverage over Somalia.[2]

Drone operations from Arba Minch were contingent on extensive logistics networks and the diverse labor of military and civilian workers. If we focus on the people, technologies, and bases that enabled these flights, we would note that the flights were supported by military personnel and facilities across the globe, through a division of labor that the military refers to as "remote split operations."[3] From ground station operators and mechanics at Arba Minch to pilots and sensor operators at Cannon Air Force Base in New Mexico (in 2013 Arba Minch operations were led by the 33rd Special Operations Squadron, based out of Cannon) to teams conducting data processing, exploitation, and dissemination in a variety of locations, in total a single Reaper Combat Air Patrol of four drones requires the work of approximately 170 military personnel.[4] Drone operations are also sustained by sophisticated sensor technologies, ground control systems, surveillance and geo-intelligence software, satellite communications, and data relay stations such as the massive Ramstein Air Base in Germany, which serves as the primary conduit for data feeds from African drone bases.[5]

And then there are the civilian logistics spaces and labors that animated this small outpost of empire. Military personnel at Arba Minch received bimonthly deliveries of food from DLA contractor Seven Seas Shipchandlers, a Dubai firm that shipped containers by sea from Bahrain to Djibouti City's port, and then hauled them overland to the facility. In addition to food supplies regular fuel deliveries for the drones were provided by the French oil and gas conglomerate, Total S.A., as part of a $51 million DLA contract.[6] Contractors also worked with military personnel on site, including drone mechanics provided by the U.S. corporation AECOM.[7] The military enrolled local sites and labor as well. A 2015 life support services contract, for instance, stated that the chosen contractor, a large nearby tourist lodge, would provide "131 bed spaces, office space for the Medics, Chaplain, Defense Operations Center, gym, laundry, internet, space to host a closed circuit television (CCTV), as well as NIPRNET [a U.S. military network for unclassified data] access/operations."[8] In addition to staff required to house and feed this contingent, the lodge was expected to supply guards for hotel security. Further afield, it is likely that analysis of drone data was provided by employees at one or more of the Pentagon's favored intelligence contracting firms in the Washington, DC, area.[9] In sum, the work required to sustain Arba Minch was remarkably extensive given that the base hosted only two drones.

The above example highlights two central elements in the support of overseas wars. The first is *logistics space*, which I define as the networked spaces of supply and support, including infrastructure, sites, equipment, information, and technologies that ensure the flow and maintenance of military people and goods.[10] Of pivotal importance here is the U.S. "global supply archipelago" of facilities located in dozens of countries across the world, including the various bases constructed and maintained in the Middle East, Afghanistan, and multiple African countries.[11] Military operations also depend on civilian logistics infrastructure, from ports and warehouses to rail and road networks to border crossings and airports. Echoing Deborah Cowen's observation about manufacturing occurring across logistics space rather than at a single site, it is not inaccurate to describe war as taking place along and through a global network of logistics spaces, not just on the battlefield.[12]

The second element—which brings the entire supply network to life—is *logistics labor*. As outlined in chapter 2, this labor is performed by an assemblage of contracting firms employing thousands of people from around the world. This is even the case with the military's most important logistics entity, U.S. Transportation Command (TRANSCOM). TRANSCOM coordinates the military's global transportation system, moving a staggering amount of goods and people around the world by sea, air, and land. In just a single year (from October 2011 to September 2012), for instance, it conducted more than 31,000 airlift missions, transporting more than 650,000 short tons of cargo and nearly 1.9 million passengers.[13] While airlift is critical, especially for movement of personnel, more than 90 percent of goods are transported by sea. In 2011 TRANSCOM's sealift branch, Military Sealift Command (MSC), estimated that "since the start of operations in Iraq and Afghanistan, MSC ships have delivered nearly 110 million square feet of combat cargo, enough to fill a supply train stretching from New York City to Los Angeles. MSC ships have also delivered more than 15 billion gallons of fuel—enough to fill a lake 1 mile in diameter and 95 feet deep."[14] Mirroring trends across the armed forces, TRANSCOM now relies heavily on contractors. In 2016 it estimated that commercial entities provided 90 percent of surface transportation (truck and rail), 55 percent of sealift support, 30 percent of airlift cargo, and 80 percent of airlift passenger transport for worldwide contingency operations.[15]

The wars in the Middle East and Afghanistan, and counterterrorism operations in Africa, have each involved distinctive combinations of logistics spaces and labor, shaped by geopolitics, physical geography, emergent wartime conditions, and preexisting economic relations and infrastructure. This has produced, in turn, different supply challenges and operational characteristics. Indeed, while the goal of logisticians is to ensure the smooth flow of people and goods, disturbances and constraints always lurk, especially in the realm of military logistics.[16] In Iraq and Afghanistan, for example, insurgents frequently targeted truck convoys and FOBs. In addition to causing supply disruptions, this also compelled, as discussed in the introduction, labor-exporting states to impose bans on citizens traveling to these countries to work for military contractors. These dynamics are absent in counterterrorism operations in Africa, where the greatest challenges involve distance and rudimentary logistics infrastructures. Consequently the military has been much more reliant on airlift to support "small footprint" operations. It is now time to examine the logistics spaces and labors of military operations in Iraq, Afghanistan, and Africa.

Iraq: Building, Maintaining, and Sustaining Baseworld

In 2005 an incredulous *New York Times* reporter wrote a story about life at Camp Liberty, part of Victory Base Complex surrounding the international airport in Baghdad. Contrary to expectations of basic amenities and oppressive heat the base, he asserted, had "the vague feel of a college campus" where troops lived in air-conditioned trailers, surfed the internet during off hours, worked out in gyms with modern equipment and a variety of exercise classes, and ate at dining halls that offered "a vast selection of food and beverages, ethnic cuisine nights, an ice cream parlor and, occasionally, a live jazz combo."[17] Four years later the paper would feature another story about bases in the country, noting that while a part of the Iraqi landscape they were in many ways "a world apart from Iraq with working lights, proper sanitation, clean streets and . . . thousands of contractors and third-county citizens to keep them running."[18]

The scale of the military's base network in Iraq at the height of operations was enormous. So too was the logistics labor required to construct and maintain it. Due to the dependence of the former on the latter, the geographical distribution of contractors is useful for limning the military's presence. This is especially the case with LOGCAP workers, whether employed by KBR or one of its many subcontractors. As noted in chapter 2, LOGCAP personnel constituted the largest portion of contractors in Iraq in 2008 (37 percent) when the number of troops in the county reached its peak. The reason for this is that through LOGCAP the military could contract KBR to conduct an incredibly wide range of services (table 5.1). Core LOGCAP tasks involved base support activities such as laundry, food, billeting, morale, MWR, facilities management, waste and sewage disposal,

TABLE 5.1. Logistics services provided by KBR in Iraq through LOGCAP III contract

Base life support services	Supply operations and material management (not procurement)	Other operations and services
Facilities management	Class I: Subsistence (food and water)	Engineering and construction projects
Laundry services	Class II: Clothing, administrative and housekeeping supplies, individual equipment (weapons, tents, tool kits, communications gear, etc)	Transportation (movement control, cargo transfer, port/terminal operations, motor pool operations and maintenance, line haul, etc)
Food services	Class III: Petroleum, oil and lubricants (POL)	Mortuary affairs
Morale, welfare and recreation (MWR)	Class IV: Construction materials	Retrograde operations
Vector and pest management services	Class V: Ammunition	Postal operations
Hazardous material storage	Class VI: Personal demand items (soap, toothpast, snacks, beverages, cigarettes, personal electronics, batteries, etc)	Ice production
Power generation and electrical distribution	Class VII: Major items (missle systems, helicopters, tanks, other vehicles, mobile machine shops, etc)	Medical services
Billetting	Class VIII: Medical supplies	Test, measurement and diagnostic equipment (TMDE) services
Water production	Class IX: Repair parts	
Waste and sewage management		
Firefighting and fire protection		
Clothing exchange and repair		
Personnel support (badging, etc)		

pest control, firefighting services, and water and power production. KBR also frequently provided materials management and operations support (but not procurement) for the military's various supply classes. This involved activities such as tracking materials, operating warehouses, and managing bulk fuel distribution. In addition to this the company performed a number of other services, from engineering and construction to transportation, ice production, and mortuary affairs support.

So what does the military's baseworld in Iraq look like from the perspective of logistics labor? Drawing on data from the 2nd quarter 2008 contractor census, figure 5.1 shows bases according to the size and composition of the LOGCAP workforce. Concerning the former I have divided the bases into four tiers, based on the number of workers. LOGCAP contingents varied substantially, from just thirteen people at a small FOB called McHenry near the town of Hawija to more than 8,700 workers at Victory Base Complex. The smallest tier of sites with fewer than 100 LOGCAP employees, like McHenry, were often FOBs with 1,000 or

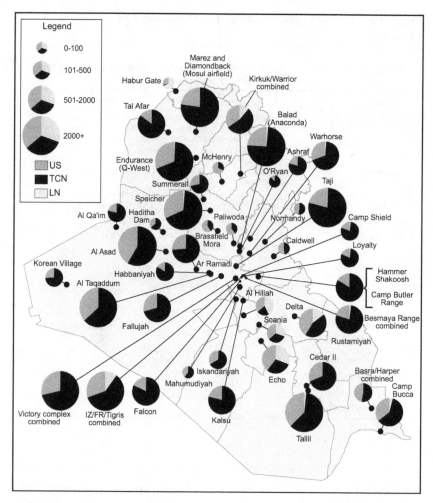

FIGURE 5.1. Distribution and composition of LOGCAP labor in Iraq during peak of military operations in 2008

fewer troops. At the other end of the spectrum the largest tier of bases with more than 2,000 LOGCAP workers were all, with the exception of the International Zone (IZ) complex, situated around airfields used for flight operations.[19] The majority of these large bases also served as key logistics hubs (which I discuss below).

At most bases in Iraq TCNs constituted the majority of LOGCAP laborers, a pattern produced in large part by KBR's reliance on subcontractors from Turkey and Gulf states, whose workers were almost exclusively recruited from South and Southeast Asia. Several thousand additional TCNs—mainly from Bosnia, Macedonia, and Kosovo—worked directly for KBR, scattered among bases across the country. Exceptions to the prevalence of foreign labor fell along two lines: 1) small bases where a handful of KBR employees from the U.S. represented the bulk of LOGCAP support personnel (i.e., McHenry, Brassfield Mora, Caldwell, and Pali-woda), and 2) bases located in predominately Shia-populated portions of the country (i.e., Al Hillah, Echo, Delta, and Scania), where KBR subcontracted a significant portion of its work to local companies that used Iraqi LN labor. In contrast, bases located in predominately Sunni-inhabited sections of the country had either only a handful of Iraqi LOGCAP employees or none at all. This does not mean that LNs did not work at these bases, just that those who did were typically contracted directly by the military through the JCC-I/A framework. For instance, in Balad—the second-largest base in the country by number of LOGCAP workers with nearly 8,200—more than 1,100 LNs provided base support and construction services through JCC-I/A contracts. The remaining outlier among LOGCAP sites when it comes to the composition of labor was Harbur Gate, the primary crossing point for goods between Turkey and Iraq, where a small contingent of KBR employees and a local company split duties.

The sprawling baseworld in Iraq did not come ready-made. Every facility, from small FOBs to large air bases with tens of thousands of troops, was the product of massive construction investments and labor. As Tom Englehart observed in 2009, the country was "a Pentagon construction site."[20] One journalist working for the DoD's own newspaper, *Stars & Stripes*, reported that by 2010 the military had spent more than $2.1 billion dollars on base construction projects in Iraq since 2004, with plans for an additional $323 million in projects to be completed before withdrawing at the end of 2011.[21] As astounding as this figure sounds, it is likely on the low side. For instance, at just a single base—Balad—construction projects worth more than $240 million were completed, initiated, or allocated funds between the beginning of 2004 and September 2005 according to a now declassified, but heavily redacted, base master plan.[22] This included $11.8 million for a wastewater treatment plant, $12.6 million for an aviation maintenance facility, $23.8 million for hospital construction and a class VIII warehouse, $7.4 million for a postal distribution center, $25 million for a fixed-wing hanger, $2.3 million for a new Army and Air Force Exchange Service (AAFES) shopping center . . . and the list goes on. In total there were nearly thirty projects in this less-than-two-year period. Two years later the base continued to be a hive of activity, a "giant construction project, with new roads, sidewalks, and structures going up" everywhere.[23]

In addition to constructing a new baseworld, by the time the number of uniformed personnel in Iraq reached its apex in 2008 the military had developed an extensive supply network to sustain operations. The linchpin holding this together was Kuwait. How important was the country for both Iraq and wider operations in the region? One indication is that over 80 percent of U.S. military forces transited Kuwait while rotating in and out of CENTCOM, with roughly 1,750,000 troops passing through in 2008 alone.[24] Similarly, the majority of supplies—from equipment to food to fuel—entered and exited Iraq from Kuwait. A remarkable 2009 DoS cable titled "A Big Footprint in the Sand: The U.S. Presence in Kuwait" details the multifaceted role this "indispensable ally" played.[25] The document begins by noting that the U.S. received over $1.2 billion annually in benefits such as "free access to bases, waived port and air support fees, customs waivers, subsidized fuel and other services." Bases that Kuwait offered "essentially open access" to the military included Ali Al Salem Air Base (the primary airport for moving U.S. forces to forward deployed sites across CENTCOM), Camp Buehring and the surrounding Udairi Range facility (used for "spin-up" or predeployment training before heading to Iraq), Camp Virginia (the main staging site for military convoys to Iraq), and Camp Arifjan, (the largest surface logistics center in the country and home to nearly 5,000 contractors). The military also had access to Kuwait Navy Base and Shuaiba, a large industrial port south of Kuwait City, while the DLA's prime food delivery contractor for Iraq, the Kuwaiti firm PWC (renamed Agility in 2006), utilized the country's largest commercial port, Shuwaikh (figure 5.2).

One of the more extraordinary elements of Kuwaiti logistical support involved border crossings, where the U.S. was given nearly unlimited control over the flow of people and supplies. One example of this, as discussed in the introduction, was Kuwait's decision not to enforce travel bans to Iraq imposed by labor -exporting countries in 2004. Another is the development of border crossings exclusively dedicated to the transit of goods and equipment by the U.S. military, coalition partners, and military contractors. The first of these, Navistar, was built next to Al-Abdali, the primary civilian border facility between Iraq and Kuwait. In 2005 Kuwait signed a memorandum of understanding (MOU) that acknowledged the "trust placed in the United States by Kuwait for day-to-day management of the Coalition Forces Crossing [Navistar]."[26] The MOU also announced plans to build a new dedicated military crossing in the desert expanse several dozen kilometers to the west, which began operations in 2007.

This new facility, called Khabari or K-Crossing, had several advantages from the military's perspective. First, all northbound military and contractor convoys were allowed to stage at Camp Virginia and other bases in Kuwait and then pass through Khabari without processing by Kuwaiti border authorities.[27] Second, there was no Iraqi government presence in the vicinity of Khabari, further facilitating the unimpeded flow of supplies.[28] Third, the new route was shorter and safer. The Navistar crossing and main supply route (MSR) arcing through south-

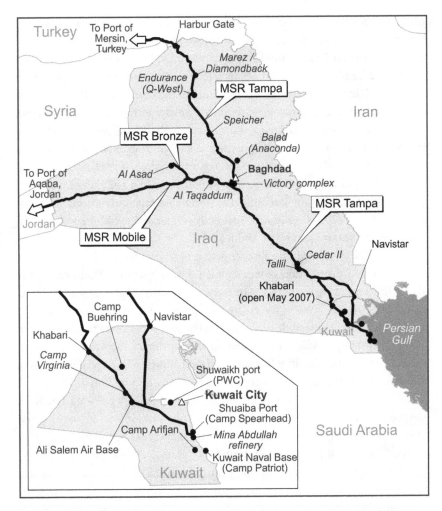

FIGURE 5.2. Iraq-Kuwait logistics network (key fuel and food supply hubs and facilities in italics)

ern Iraq passed through several towns prior to reaching the first two logistics hubs south of Nasiriyah (Camp Cedar II and Tallil Air Base), increasing the risk of improvised explosive device (IED) attacks and hijackings. In contrast, the new route from Khabari to these bases was a relatively straight shot through desert with "virtually no habitation."[29]

Another important aspect of Kuwaiti support involved fuel. From late 2002 to March 2005, the country supplied aviation fuel to the military free of charge and from 2005 to the end of 2008 it sold fuel at below-market rates.[30] It also approved the military's prewar construction of a fuel pipeline that ran from Mina

Abdullah refinery in the southeast corner of the country to Camp Virginia, and then forward to a pump station at the border in the northwest of the country, not far from where Khabari would eventually be built. Just hours after the invasion commenced, military engineers began extending the pipeline, called the Inland Petroleum Distribution System (IPDS), through the Iraqi desert. IPDS consisted of thousands of nineteen-foot-long, six-inch-diameter sections of aluminum pipe, joined by coupling clamps. Due to aluminum's high thermal reactivity, changes in temperature could cause the pipeline to shrink or expand by two feet for every fifty sections, thus expansion loops were required at regular intervals, as were smaller pump stations every twenty kilometers. In a remarkable display of engineering capability, by late April the completed pipeline—with a through-put capacity of 720,000 gallons of fuel daily—extended to Tallil and Cedar II, 360 kilometers away from Mina Abdullah.[31]

As the military settled into its occupation of Iraq, it required a more robust and flexible fuel distribution system than IPDS, which was a tactical solution designed for use during the outset of operations. By early 2009 nearly 1,500 DLA-contracted trucks a day were traversing the region delivering gasoline, diesel, and aviation fuel from Kuwait, Jordan, and Turkey to a series of large bases located along the military's MSR network that served as primary fuel storage sites (figure 5.2). From these facilities fuel was then distributed to the rest of the bases in the country. In total the military consumed more than 1.5 million gallons of fuel a day, the vast majority of this being jet fuel.[32] More than 60 percent came from suppliers in Kuwait. DLA also shipped fuel to ports in Jordan (20 percent) and Turkey (13 percent), where it was then loaded onto trucks bound for Iraq.[33] Paradoxically only a small fraction of the military's fuel needs was provided by suppliers in Iraq, even though the country possesses some of the largest oil reserves in the world. In one of history's many ironies, U.S. dependence on fuel imports to sustain operations stemmed from the collapse of Iraq's refining capabilities due to infrastructure damage during Operation Desert Storm and the imposition of sanctions in the decade following.

Afghanistan: The Geopolitics of Supplying a Logistics Island

Perhaps the defining difference between the Iraq and Afghanistan campaigns concerns the spaces and geopolitics of logistical support.[34] In the former an extensive road network facilitated the flow of goods, and preexisting Iraqi military installations—most notably large air bases situated throughout the country—could be developed to serve as logistics hubs. Even more important was Kuwait's willingness to serve as the primary staging area for personnel and material, enabling use of its modern port facilities and airfields, spin-up training at desert bases, U.S. control over border operations, and reliable provision of refined fuel products. In addition to Kuwait, goods to Iraq could be routed through Turkey and Jordan, both U.S. allies. The military's supply chain for operations in Iraq, in sum, was relatively short and robust.

Afghanistan, in contrast, is more akin, as Pierre Belanger and Alexander Arroyo have observed, to a "logistics island."[35] To begin it is a landlocked country. One of the closest ocean-going ports, Chabahar in Iran, is more than 900 kilometers from the nearest Iran-Afghanistan border crossing. For geopolitical reasons, Iran is not a viable option for transiting U.S. or NATO coalition military supplies into Afghanistan. Instead, for much of the war the military's primary ground line of communication (GLOC) has run through Pakistan. This is an incredibly long supply route with two branches (figure 5.3). The first extends from the port of Karachi, to Quetta and the Chaman border crossing, then on to the massive base at Kandahar Airfield (KAF), the military's main logistics depot in southern Afghanistan roughly 900 kilometers away. The second, longer— approximately 1,700 to 2,000 kilometers depending on the route taken through Pakistan—and more perilous branch to Bagram Air Base, the primary logistics center in the north of the country, runs from Karachi to Peshawar, through the Torkham border crossing, and then over Khyber Pass into Afghanistan. Transit of goods from Karachi to Kandahar or Bagram typically takes one to three weeks. But accidents, strikes, and delays at the border crossings have often produced significant delays, forcing the military "to budget months for travel that should take days."[36]

In addition to long supply lines and frequent delays, utilizing Pakistan as the primary logistical conduit to Afghanistan has presented several other limitations and deficiencies for the U.S. military compared with operations in Iraq. First, due to Pakistani government restrictions, troops, weapons, and ammunition must be flown to bases in Afghanistan, whereas in Iraq the military had the option of staging troops and equipment in Kuwait and then traveling overland to bases. Second, the GLOC through Pakistan is much less secure than routes through Kuwait, Turkey, or Jordan, with supply operations in the former plagued by pilferage and attacks on trucks, bridges, and staging areas. Part of the problem is the inability of U.S. personnel to oversee the flow of goods. As AMC's deputy commander acknowledged in 2010, "Once the piece of equipment gets off the boat at Karachi, no American [soldier] touches it—it is all contract [labor] because of the political situation in Pakistan."[37] Consequently the military has increasingly turned to remote technologies such as radio-frequency identification tags, shipping container intrusion monitoring devices, and satellite tracking to combat the

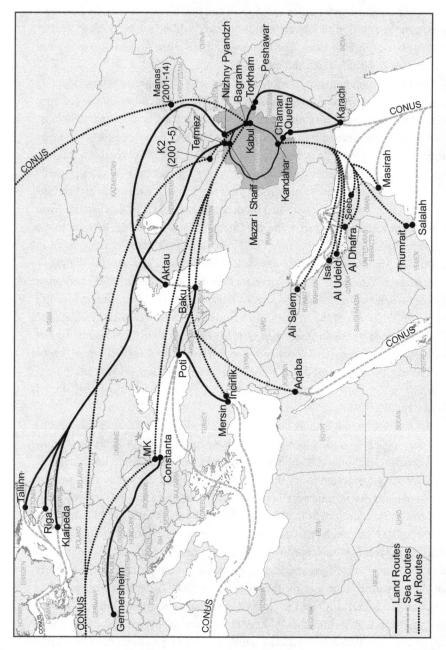

FIGURE 5.3. Main supply lines and logistics sites for Afghanistan operations (lines are approximate and do not indicate exact routes taken)

problem of en route theft.[38] A bigger issue has been the inability of Pakistan to prevent militants from carrying out attacks within its territory, a problem that was especially acute around Peshawar and the Torkham crossing in 2008–9. In December 2008, for instance, around 300 cargo trucks and military vehicles were destroyed in a series of attacks on staging yards in Peshawar.[39]

The most significant challenge, however, has been the fraught geopolitical relationship between Pakistan and the U.S. Though nominally allies, contradictory interests and mutual distrust pervade the relationship, a tension colorfully captured by one senior U.S. diplomat's characterization of the two countries as "frenemies."[40] Indeed, despite public praise by U.S. officials calling Pakistan a crucial partner for counterterrorism and counterinsurgency campaigns in the years after 9/11, behind the scenes there is deep concern that the country supports, or at least tolerates, the activities of various extremist groups within its borders who in turn carry out attacks against U.S. forces in Afghanistan. Following the 2011 raid on Osama bin Laden's compound in Abbottabad—and reports that he had been hiding out in this military town located only fifty kilometers from Pakistan's capital for years—such suspicions swelled. Pakistani political and military authorities in turn have their own concerns, including fear that the U.S. is increasingly orienting itself toward their main rival, India.[41] Another irritant from Pakistan's point of view are drone strikes—and occasional cross-border raids—conducted primarily in the country's northwest borderlands, the Federally Administered Tribal Areas (FATA). Drone strikes began in 2004, peaked in 2010, and continue to the present day. In total more than 400 strikes have been conducted killing up to 4,000 people, nearly a quarter of them civilians.[42]

In 2008 the U.S. began to put into motion long-standing plans to develop the Northern Distribution Network (NDN) for Afghanistan, with the goal of reducing reliance on the Pakistani supply route. This was not a completely new concept. As far back as 2005 DLA had begun sourcing fuel from Azerbaijan and Kazakhstan, with roughly 30 percent coming from these two countries by the end of 2007. Additionally, prior to 2005 TRANSCOM shipped some food and construction materials across Europe to Afghanistan by rail.[43] It also used air bases in Uzbekistan (Karshi-Khanabad, known as K2, 2001–5) and Kyrgyzstan (Manas, 2001–2014) as transit centers for flying personnel into Afghanistan.[44] In addition to insecurity in Pakistan the NDN initiative was motivated by two further calculations. First, troop levels in Afghanistan were growing—even before the Afghanistan surge implemented by President Obama in 2009—which led to worries about a lack of surplus capacity along the Pakistan route.[45] Second, U.S. officials were concerned that Pakistan might threaten to close the border as geopolitical leverage, or in response to cross-border operations. This was not a theoretical concern as Pakistan closed the border for several days in 2008 after the bombing of a military outpost in FATA.[46]

The NDN was originally constituted by two distinct routes, each with their own variants (figure 5.3). The first "Russian" route began at the Latvian port of Riga (with operations later expanded to include Tallinn and Klaipeda in Estonia and Lithuania) where cargo was loaded onto trains. After traversing Russia, Kazakhstan, and Uzbekistan by rail, goods were then unloaded at the Uzbek border town of Termez and then driven to their final destinations in Afghanistan. Alternatively cargo would be unloaded in Kazakhstan and then hauled by truck though Kyrgyzstan and Tajikistan and then into Afghanistan at the Nizhny Pyandzh border crossing. In 2015 Russia rescinded transit permission across its territory, thereby closing off this northern circuit. The second "Caucasus" route begins at the Georgian port of Poti. After crossing Georgia and Azerbaijan by rail, cargo is then ferried across the Caspian to Kazakhstan, where it is then reloaded onto trains that end at Termez. As with the Russian route a variant of this approach utilizes line haul across Kazakhstan, Kyrgyzstan, and Tajikistan into Afghanistan.[47] In 2010 the military also began shipping goods through the port of Mersin in Turkey, even using overland transport from DLA warehouses in Germersheim, Germany, to feed into this second supply line.[48]

The first NDN shipments began in 2009, and by June 2011 nearly 40 percent of supplies to Afghanistan were being delivered via the network despite the fact that this was nearly three times more expensive than routing cargo through Pakistan.[49] U.S. reliance on the NDN routes increased dramatically later that year in the wake of the decision by Pakistan to close its borders to Afghanistan-bound military supplies for eight months following a U.S. military attack on two border posts that resulted in the deaths of twenty-four Pakistani soldiers in November. The military also relied upon the NDN to facilitate retrograde—military speak for removal—of equipment during troop withdrawals beginning in 2012, while at the same time increasingly utilizing airlift to move equipment to sites in the Middle East and Europe, where it is subsequently loaded onto ships for delivery back to the U.S.[50]

Any analysis of the NDN needs to go beyond a narrow economic calculation of shipping costs as its development has also affected politics and human rights in the region. The linchpin through which the vast majority of cargo enters Afghanistan, for instance, is Uzbekistan, an authoritarian state with one of the worst human rights records in the world.[51] Or as it was put more delicately by the U.S. embassy in Tashkent in 2009 when it was cultivating Uzbek support for the NDN scheme, "A non-democratic regime with a troublesome human rights record in the center of a strategically important, but unstable region."[52] It is also a country that is extremely sensitive to criticism along these lines. Beginning in late 2001

Uzbekistan granted the U.S. use of an air base at K2 in the south of the country. In exchange the U.S. provided more than $200 million in military hardware and surveillance equipment the following year.[53] Uzbekistan also received support for its campaign against the Islamic Movement of Uzbekistan, a group of radical militants that were also fighting alongside the Taliban in Afghanistan.

This geopolitical quid pro quo was not to last. In May 2005 Uzbek security forces killed several hundred protestors in the southern city of Andijan, leading to calls for an international investigation by human rights groups, DoS, and a number of U.S. senators. Of particular concern for critics were reports that military hardware provided by the U.S. was used in the attack.[54] Angered by these criticisms and fearful that Andijan might be used as a pretext for a "color revolution"—especially following the collapse of the ruling regime in neighboring Kyrgyzstan just months before in the Tulip Revolution—Uzbekistan's president, Islam Karimov, moved quickly to evict the U.S. from K2. The realpolitik lesson U.S. officials learned was "to not push Central Asian regimes too hard on democracy and human rights issues, especially when important security cooperation and basing rights were at stake."[55] It was a lesson they would not forget during subsequent NDN operations, leading to accusations that they were "whitewashing . . . abuses" of the Karimov regime and other states in the region.[56]

Arguably the most striking difference between Iraq and Afghanistan concerns the practice and geopolitics of logistics operations *within* each country. Consider the distribution process for food supplies to bases in the two countries in 2008, which was supervised by DLA. In the case of Iraq military escorts would meet truck convoys at the Kuwaiti border (Khabari) and then travel with them to one of the primary logistics centers in Iraq (figure 5.3). From there truckers would unload their goods or pick up new escorts that would travel with them to their final destination. After deliveries were completed the trucks were then escorted back to Kuwait, where the process would begin again. In Afghanistan, in contrast, trucks carrying DLA foodstuffs were not provided with military escorts, whether hauling supplies to the two primary logistics hubs in the country (Bagram and Kandahar) or delivering goods directly to FOBs. Also, rather than entering immediately trucks were required to stage outside a FOB for at least twenty-four hours in a "cooling yard" where they were inspected for IEDs by contractors or Afghan National Army personnel. Upon completion of delivery, truckers then returned to supply warehouses, again unescorted.[57] As with food, DLA deliveries of fuel in Iraq were accompanied by military escorts at all times while military escorts were generally not provided for fuel trucks in Afghanistan.[58] A 2009 Host Nation Trucking (HNT) contract that simplified supply operations by awarding contracts to six prime trucking contractors in Afghanistan codified this practice by stating that the "contractor is responsible for all security" and convoys should be conducted "independently, without military escorts, unless otherwise determined by the USG [U.S. government] at its sole direction."[59]

Unprotected logistics supply lines, of course, are inviting targets for any enemy. Truck convoys in Afghanistan have been frequently attacked—and with deadly results. According to one military briefing, attacks on convoys resulted in nearly 100 fatalities between December 2005 and February 2008.[60] As supply operations expanded the following year in response to the troop surge, attacks and casualties mounted. Trucking companies reacted by turning to private security companies to protect their convoys, a strategy that was mandated by the 2009 HNT contract. In practice, however, private security companies are often little more than thinly disguised fronts for local warlords who run what amounts to a protection racket, demanding bribes in exchange for refraining from attacking trucks that transit territories they control.[61]

Even more concerning is ample evidence that a significant portion of the fees earned by Afghan security companies have ended up in the pockets of the Taliban and other insurgents, perhaps as much as 10 percent of logistics contracts—amounting to hundreds of millions of dollars—according to an explosive report by *The Nation*'s Aram Roston in 2009.[62] As one U.S. contractor in this article observed, "the Army is basically paying the Taliban not to shoot at them." Pakistan's subsequent closure of the border would demonstrate that this quote was not an exaggeration—and that the practice of paying the Taliban not to attack truck convoys is not limited to Afghanistan. Following the reopening of the border in July 2012, a series of news accounts suggested that the Taliban was more adversely affected by the suspension of the Pakistan GLOC than the U.S. military. According to one Taliban commander, "The NATO supply [route] is very important for us," and in fact, "stopping these supplies caused us real trouble" as "earnings dropped down pretty badly. Therefore the rebellion [in past months] was not as strong as we had planned."[63]

Two years later John Sopko, head of the Special Inspector General for Afghanistan Reconstruction, the U.S. government watchdog for military and civilian reconstruction efforts in Afghanistan, sharply castigated military officials for their inaction in relation to this problem:

> As I have pointed out in our last six quarterly reports, the Army's refusal to suspend or debar supporters of the insurgency from receiving government contracts because the information supporting these recommendations is classified is not only legally wrong, but contrary to sound policy and national-security goals. I remain troubled by the fact that our government can and does use classified information to arrest, detain, and even kill individuals linked to the insurgency in Afghanistan, but appar-

ently refuses to use the same classified information to deny those same individuals their right to obtain contracts with the U.S. government. There is no logic to this continuing disparity.[64]

Actually there was a logic. It was just an insidious one. Several Afghan security contractors suspected of funneling protection payments to the Taliban were politically connected and ostensibly coalition allies. Among those highlighted in Roston's article were Watan Risk Management, which was run by two cousins of Afghanistan's then-president, Hamid Karzai, and NCL Holdings, run by the son of the then-defense minister, Abdul Rahim Wardak.

The irony of supply operations in Afghanistan providing a key source of funding for insurgents waging war against U.S. forces is that the decision not to provide military escorts was motivated in part by a concern for casualties that this would entail. In Iraq, attacks on truck convoys in 2004 compelled the U.S. to bolster military escorts, in part to head off travel bans by labor-exporting states like India and the Philippines. Though ultimately unsuccessful in achieving this goal, the practice continued throughout the occupation of that country. Providing military escorts, however, carries significant risks for U.S. troops. In early 2011 Steven Anderson, a senior military logistician who was in Iraq in 2006–7, estimated that approximately 1,000 U.S. personnel had been killed while on "fuel-related missions in Iraq and Afghanistan," with the bulk of casualties occurring in the former theater.[65] Remarkably, this represented nearly one-quarter of all battlefield deaths suffered in the two conflicts to that point. Another particularly evocative article in *Armed Forces Journal* the following year referred to the "direct link between fuel [demand] and casualties" as "logistical fratricide."[66]

As noted above, logistical operations in Iraq were relatively straightforward compared to those in Afghanistan. Therefore the decision not to provide military escorts for truck convoys in the latter—which was made in late 2003 or early 2004—made sense, initially.[67] In the end, however, attempts to avoid logistical fratricide have just displaced the problem and ultimately provided the monetary fuel for insurgent operations across the country. Logistics contracting in Afghanistan amply illustrates, in others words, Derek Gregory's argument that "the business of supplying war produces volatile and violent spaces in which—and through which—the geopolitical and geo-economic are still locked in a deadly embrace."[68]

Africa: Developing the Sinews of Support for Counterterrorism Operations

In contrast to bases in the Middle East and Afghanistan facilities in Africa tend to be small and austere. At one level this reflects the military's preference for a small footprint approach to deployment in Africa Command (AFRICOM) that eschews large bases in favor of a network of small, relatively unobtrusive "lily pads" that facilitate force projection across the continent.[69] But as the following vignette from a Navy SEAL who served as commander of several dozen operators at Camp Simba in Manda Bay, Kenya, in 2005 illustrates, this also reflects the difficulty of supplying personnel in Africa: "We had no fresh fruit or vegetables at Manda Bay. Our supply officers in Djibouti tried to get us fresh fruit, but it was difficult to transport an orange from Europe to Djibouti, from Djibouti to Mombasa, and from Mombasa up to Lamu. We ate peaches soaked in syrup packaged in MRE bags."[70] Military presentations, reports, articles, and theses on Africa almost inevitably include comments about the logistical challenges that operations pose due to the "tyranny of distance" and underdeveloped transportation infrastructure on the continent.[71] Consequently, in recent years the U.S. military has focused attention on the development of more robust logistics networks, with an assemblage of contractors playing a key role in providing services.

Contracting facilitates two related operational priorities on the continent. First, it allows the military to maintain a relatively low profile, even as it operates from dozens of facilities, including several drone bases and SOF compounds (figure 5.4).[72] Due to Africa's vast size, contracted air transportation is key, and is of particular importance for SOF teams who conduct operations across a wide swath of the Sahel, Maghreb, and Central and East Africa, and who rely on U.S.-based flight contractors. In 2016, for instance, Special Operations Command, Africa (SOCAFRICA) issued a solicitation for two helicopters based out of a previously unknown base in Arlit, Niger, to provide support for military operations in the "North and West Africa Area of Operations," which includes the countries of Mali, Algeria, Libya, Tunisia, Chad, Cameroon, Nigeria, Benin, Mauritania, Senegal, and Burkina Faso. Notably the contract language specified that "aircraft shall not be painted in a color that is close to military colors and paint schemes. A conservative, predominately white, civilian-style paint scheme is preferred."[73] Contracting documents indicate that the main hubs for fixed- and rotary-wing air transportation for SOF operations include Entebbe in Uganda, and Niamey in Niger.

Airlift works well for small, mobile SOF teams. But as the military expands its presence across the continent, the cost and limitations of transporting goods and equipment by air has precipitated efforts to develop "adaptive" logistics networks that utilize international and African surface transportation firms. In 2011 one

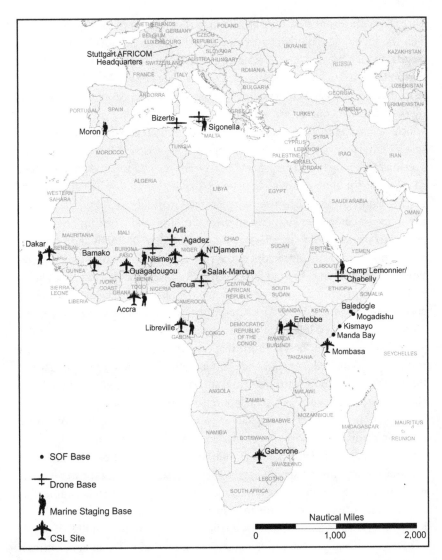

FIGURE 5.4. Drone bases, CSLs, Marine staging bases, and SOF sites supporting operations in Africa

of the first such initiatives, dubbed "the new spice route," combined sealift with line haul by local truck companies to move goods between several bases and temporary forward operating locations (FOLs) in Kenya, Uganda, Ethiopia, and Djibouti.[74] Following the successful completion of this operation the aforementioned Dubai firm Seven Seas Shipchandlers was awarded a two-year contract by the DLA to make twice-monthly refrigerated and dry goods truck deliveries to three facilities in the "Ethiopia Operational Deployment Zone": Camp Gilbert at

Dire Dawa, the drone base at Arba Minch, and an unknown facility in Negele.[75] This was followed by another two-year contract in 2013. Nothing is known about Negele, though it is possible to infer from the contract that a similar number of personnel (~100+) were located there as there were at Arba Minch, which suggests that it was a key SOF base for missions in Somalia. Seven Seas Shipchandlers and its subcontractors also regularly deliver supplies to Camp Simba, which, according to 2016 contracting documents, has expanded into a facility that hosts a steady state population of "approximately 325 military personnel with potential surges up to 510 personnel."[76] In recent years AFRICOM has developed a surface distribution contracting network that extends across the entire continent, beginning with a 2014 award for contracts worth up to $10 million each to five different companies "to perform surface transport and distribution of general cargo within all fifty five (55) nations of the AFRICOM AOR and Egypt."[77]

The use of civilian contractors to the reduce the visibility of military operations in Africa extends beyond logistics to ISR as well, the best examples being two previous manned surveillance operations—codenamed Creeksand and Tuskersand—in West and Central Africa, respectively.[78] As with air and ground transportation, contracted ISR operations are by design intended to be as low-profile as possible to reassure "host nations" that are "uncomfortable with U.S. military platforms."[79] Tender documents and contracts often include specific language on the number of personnel, the flight and surveillance equipment to be used, and aircraft appearance, such as a solicitation from 2010 that states that operations must "present a relatively inconspicuous presentation, including but not limited to: (i) no distinctive US or military markings, other than the required US registration number and placards, and (ii) no 'one of a kind' platform which would invite attention. Aircraft should have a 'slick' appearance with little to no external variation (i.e., antenna arrays, baggage pods, fuel pods)."[80] Small, unmarked, civilian planes favored by contractors, such as the Pilatus PC-12 and Beechcraft King Air series, also have the added benefit of requiring only a handful of people to operate and being capable of flying out of remote and rudimentary airfields if necessary.

In addition to an unobtrusive presence logistics contracting also facilitates AFRICOM's stated goal of organizing force posture "to maximize operational flexibility and agility."[81] Contractors provide base operations and life support services to the growing number of U.S. military sites and operations in Africa. In some cases—such as SOF facilities and drone operations in Niger, Cameroon, and Somalia; small bases in Uganda, Central Africa Republic, Democratic Republic of the Congo, and South Sudan that were used in counter-Lord's Revolutionary Army operations; and OUA—these contracts are awarded through established channels such as LOGCAP.[82] In other instances ad hoc solicitations or no-bid contracts are used due to the small size of the base, temporary duration of operations, or difficulty in identifying qualified providers. In 2015, for instance, the Marine Corps solicited bids to provide base support services for up to four months for twenty-four troops conducting training exercises with the Ugandan military at Camp Singo in Uganda in the fall.[83] The U.S. frequently uses Camp Singo—which is located approximately seventy kilometers northwest of Kampala—for training exercises with Ugandan and other African military contingents, and has even established a small fenced compound with buildings, tents, water tanks, and generators.[84] But rather than permanently stationing troops there, it rotates them in as desired, relying on short-term contracts for base and life support.

The development of contracting capabilities in Africa has been accompanied by the emergence of an increasingly dense network of sites that facilitate the movement of U.S. personnel and equipment across the continent. Foremost among these are Cooperative Security Locations (CSLs), which the DoD characterizes as facilities "with little or no permanent U.S. presence, maintained with periodic Service, contractor, or host nation support."[85] CSLs are typically located at large airports and are valued because they provide "a foothold for conducting the full range of military options, forced entry, humanitarian relief, NEO [noncombatant evacuation operation], peacemaking, peace keeping, and other stabilization operations."[86] Following the September 2012 attack on the U.S. consulate in Benghazi, Libya, that resulted in the death of U.S. ambassador Christopher Stevens, AFRICOM began upgrading several CSLs into staging bases for use by Special Purpose Marine Air-Ground Task Force, Crisis Response (SPMAGTF-CR) teams as part of a strategic shift that military documents and online resumes refer to as the "new normal" or Operation New Normal.[87] These expanded CSLs are, according to one military article, capable of hosting "within hours . . . nearly 200 troops for as long as they need to stay."[88] News accounts and contracting documents suggest that SPMAGTF-CR bases exist or are being set up in Ghana, Gabon, Senegal, Niger, and Uganda, with SPMAGTF-CR units also operating out of larger military bases in Djibouti, Spain, and Italy (figure 5.4).[89] In addition to CSLs and Marine staging bases, the Navy also utilizes a number of African ports for fuel bunkering. Moreover, all of these sites are supported by an extensive network of logistics nodes across Europe and the Middle East, and existing strategic airlift and sealift channels and sites maintained by TRANSCOM.[90]

AFRICOM's growing logistics network, in conjunction with an increasingly robust assemblage of contractors, ranging from small African trucking firms to massive multinational corporations, supports more than 1,700 SOF troops on the continent.[91] It also facilitates a remarkably large number of military actions, from joint training exercises with dozens of African and European militaries (12 in 2015), to security cooperation activities (400 in 2015), to military operations

(75 in 2015), all while promoting flexibility and a low-on-the-ground profile that belies this activity.[92] This in turn has deepened political and military entanglements between the U.S. and African governments, especially in the realm of counterterrorism activities.

Similar to the case of securing logistical support in Central Asia for operations in Afghanistan, the political effects of these entanglements have often been deleterious. Four of the countries that AFRICOM has established the closest counterterrorism partnerships with have experienced successful or attempted military coups in recent years: Mauritania (2005 and 2008), Niger (2010), Mali (2012) and Burkina Faso (2014 and 2015). In the case of Mali, the coup was led by Captain Amadou Haya Sanogo, a participant in "several" U.S. military training programs, while the leader of the most recent coup attempt in Burkina Faso, General Gilbert Diendere, was the country's "point person on the U.S. Trans-Sahara Counter Terrorism Partnership."[93] Chad, another key partner, has seen several attempted coups against an authoritarian government led by Idriss Deby, who himself came to power though a coup in 1990. As the Oxford Research Group remarked in a 2014 report, "The pursuit of counterterrorism operations and basing or logistics infrastructure across the Sahel-Sahara is dependent on maintaining relationships and status of forces agreements with national governments," with the result being that these states have become "largely immune from pressure to improve their repressive treatment of citizens and political opponents" due their status as reliable partners in the "war on terror."[94] In other words, for political and military elites in the Sahel, binding themselves to AFRICOM's counterterrorism assemblage can be useful for better securing their own authority and privileges against potential challengers. This dynamic is not limited to Africa—or Central Asia—as the U.S. has "repeatedly collaborated with murderous, antidemocratic regimes and ignored widespread evidence of human rights abuses" in countries that it relies upon for overseas bases of operation.[95]

ASSEMBLING A TRANSNATIONAL WORKFORCE

> **There were so many people in the streets. Maybe a thousand people. . . . You would register [with the recruiting agency] and then wait, because you never knew when your name would be called. If it was called and you weren't there you missed your chance because it wouldn't be called again. After one week I heard my name.**
>
> —Danilo

The previous chapter identified logistics spaces and labor as two foundational elements of military operations. While the former receives more attention, it is the latter that animates war. Whether drone flights at remote locations in Africa or counterinsurgency campaigns in the Middle East, the U.S. military depends on the beating heart of logistical labor. Due to the increase in contracting, the composition of this labor is increasingly civilian and foreign rather than American and uniformed. Consequently the military is now inextricably entangled with the business of transnational labor acquisition, as uncomfortable as it is with acknowledging this fact.

Assembling a constantly shifting workforce of hundreds of thousands of individuals from around the world is itself a massive logistical undertaking, one that involves its own distinctive combinations of sites and labor. It depends on a vast "migration infrastructure" of "systematically interlinked technologies, institutions, and actors that facilitate and condition mobility."[1] Some elements of this migration infrastructure, like recruiting agencies, government bureaucrats and websites, are well known, while others, such as hotels and suburban malls, less so.

In this and the following chapter I trace the various routes traversed by Filipinos and Bosnians who have worked in the Middle East, Afghanistan, and Africa over the past two decades. The focus of this chapter is legal, or visible, labor procurement, while the next examines trafficking and underground recruiting channels. In practice, of course, legal and illegal recruiting can be rather difficult to neatly delineate, and thus it is better view them as positions along a spectrum rather than dichotomous categories. How, for instance, should one categorize the experience of Bosnians who signed Asian contracts with DynCorp because they were falsely promised that they would be able to switch over to a European contract when they arrived in Afghanistan? Despite this deception, I include their accounts here due to the relative lack of coercion experienced by Bosnians compared to the examples of trafficking from South and Southeast Asian countries, such as the case of the twelve Nepalese workers killed in Iraq.

For both Filipino and Bosnian workers, the experience of gaining employment with logistics contractors has been greatly influenced by their countries' respective histories of involvement with the U.S. military, and the Philippines' position as a labor-exporting state, as discussed in chapters 3 and 4. This has produced distinct recruiting processes, as well as differences in the types of firms that seek labor in each country, as illustrated by the following two stories. The first, told by Carlos, is at once both serendipitous—in his telling—and indicative of the frantic atmosphere accompanying the mad dash by KBR's largest subcontractor, PPI, to amass thousands of Filipino workers in early 2004 to fulfill its contractual obligations in Iraq.

> CARLOS: I was visiting my wife in Manila. She was a secretary at a school near Anglo [AES]. And I saw many people in the streets. And I just got curious. People sleeping in the street, waiting for an opportunity. I saw when I was riding by in a jeepney. So I stopped. Because every time I go to Manila I bring my passport and résumé. Then I went to people and asked—I met friends from Pampanga—and they said, "They need workers in Iraq, salary is $600 a month."
>
> ME: So what was the interview like?
>
> CARLOS: They asked me about international cooking. How to cook a steak, how long, how to make a sauce. I passed all those questions. "OK you are hired. You can go in three days." I didn't even go back home to Pampanga, I just waited there for three days!
>
> ME: Did you tell your wife?
>
> CARLOS: I just told my wife, "Bring me some clothes, I need this and this." She said, "Why?" "Because I am going to Iraq."
>
> ME: Did you see your children?
>
> CARLOS: I didn't see my children, only my wife, because she was working in Manila. And I left my wife three months pregnant at that time. I was really lucky. Because many Filipinos were waiting a month or more, but only three days for me.

Carlos's experience can be contrasted with Asim's account of obtaining a position with KBR in 2006:

Me: How did you get the job?

Asim: I applied online actually. I applied online and one day some woman [working for KBR] called me and said, "Are you still interested in the job?" "Of course I am." She asked me if I had a passport. She asked "Can you come tomorrow in Sarajevo?" I said, "Why not?"

Me: Where in Sarajevo?

Asim: A hotel, the Holiday Inn. First I spoke with one guy. A big American guy, a bull like me. He asked simple questions. He wanted to know if I understand, you know, can we speak same English language. He asked simple questions about how I travel, what kind of car I drive, what kind of weather is outside. After that I had a conversation with three different people about the job.

Me: What kind of job did you apply for?

Asim: I was electric, electric mechanic. That was my first job down there.

Me: So construction of buildings and wiring?

Asim: Yeah. And I passed all those tests. And seven or ten days after they sent me to America, to Houston. I was in Houston for like four weeks. And after that a straight flight to Dubai, and after Dubai, Baghdad.

For Filipino workers like Carlos, obtaining a job with a subcontractor such as PPI was with few exceptions mediated by recruiting agencies, like AES, which serve as the linchpin of the labor export system established by the Philippines. In the absence of such a system, Bosnian job seekers like Asim navigate an alternative world populated by hotels and websites offering job postings or information on the recruiting practices of prime contracting firms like KBR. In the following two sections I examine further the distinct temporalities and geographies of recruiting in the two countries. The chapter concludes with a brief consideration of the logistics involved in assembling a global workforce.

Recruiting Agencies, Body Shops, and Fast Labor Acquisition

Recruiting agencies occupy a prominent place in Filipinos' accounts of obtaining a job with military logistics firms. This is even the case for many who found work in Iraq and Afghanistan after travel bans were imposed, and agencies could have their licenses revoked for working with contractors seeking labor for projects in those countries, as I discuss in the next chapter. The reason for this is that while the state regulates labor export it is the agencies that serve as labor brokers that connect foreign firms with Filipino workers.[2] In simplified form, the process proceeds as follows. First prospective employers select a recruiting agency to help them fulfill their labor needs. This is facilitated by labor niche specialization, as the thousands of firms competing for business tend to specialize in certain regions and/or occupations to increase their competitiveness.[3] After selecting an agency prospective employers then register with the POEA, which provides accreditation allowing them to seek Filipino labor. This step can either be done directly or by their chosen firm. At this point the agencies take center stage, recruiting qualified workers and processing them for foreign deployment.[4]

One significant fact about the industry is that almost all recruiting agencies are based in the Metro Manila region, typically operating out of small, nondescript office buildings. There are several reasons for this spatial agglomeration of operations. The most important is that being located in the capital next to regulating agencies and foreign embassies facilitates rapid acquisition of required government documents and overseas visas. In addition to this, certification and testing for certain occupations, such as sea-based workers and performing artists, is concentrated in Manila. The city is also host to a large number of occupational schools and universities that focus on training workers for overseas jobs. More generally, as the country's primate city Manila offers the largest potential pool of skilled and unskilled labor.[5]

Consequently, living or working in Manila, as Carlos's story above illustrates, greatly increases the chances of learning about overseas opportunities, whether through happenstance or personal connections. One example of the latter pathway is provided by Flora, who also began working for PPI in 2004: "I have always wanted to work overseas. Because I wanted to give my mother a comfortable life. I applied as a domestic helper [before] but wasn't hired. I heard about the job because the secretary of Mr. Helliwell [Neil Helliwell, the CEO of PPI] lived on the same street as me. Her sister is my childhood friend. Her sister approached me and asked if I wanted to work in Iraq. And I said, 'Yes, why not?'"

PPI was just one of several companies seeking labor in the Philippines shortly after the U.S. invasion of Iraq. Another prominent one was the Turkish firm Serka, which was awarded a subcontract by KBR for staffing and management of several DFACs at bases in northern Iraq in summer 2003. Michelle, the wife of one of the first workers hired by Serka, remembers that "in October 2003 when we were riding in a bus there was a newspaper advertisement for bakers and cooks at U.S. bases, and a good salary. When he [her husband] came back from Saudi [Arabia] we opened a bakery, but you can't make much money here. So we saw the advertisement [that said] 'Baker $800' and came to the agency, Blazing [Star]. He was in the first batch to go to Iraq. Processing was only 10 days. . . . This was the same time that Bush was here visiting Arroyo." Like several Filipino workers I interviewed, Michelle's husband had previous experience in the Gulf region, working as a chef for years in Saudi Arabia. Michelle wanted to work in Iraq too, but was told by Blazing Star that Serka did not permit couples to work in the DFACs. So instead she decided to serve as a local recruiter for the agency.

Local recruiters are another essential component of the overseas recruiting network in the Philippines. Located in villages and towns outside Manila, or on its outskirts, they work with agencies to advertise opportunities with neighbors, friends, and family. In exchange they are typically paid a fee for each person they successfully direct to an agency. A former PPI employee in Iraq who now works as a local recruiter in her village told me she receives 1,000 pesos per person, which is roughly $20. Local recruiters may also provide guidance for the application process. Michelle estimates that she helped more than 100 people get a job with Serka in the decade after her husband went to Iraq. Roughly half of these were from her *barangay* (village or neighborhood). What is remarkable about this is that few had previous experience in the food service industry, and most of the men—who constituted the majority of recruits from her neighborhood—did not even have rudimentary cooking knowledge. So Michelle devised an informal two-to-four week cooking and baking "boot camp," turning construction workers and tricycle drivers into bakers, pastry chefs, and kitchen assistants. "They didn't know anything when we started. I had to teach them the basics about flour," Michelle recalls. "I even approached bakeries here and asked them if they would let the men work without pay for a couple of weeks so they would be able to learn more about it."

In some cases local recruiters have preexisting personal or familial connections with recruiting agencies. AES enrolled family and friends in the Pampanga region northwest of Manila, where the Arcilla family is from. This involved setting up temporary satellite recruiting centers in homes according to several workers from the region. One, Sam, recalls that "the recruiter was from my *barangay*, Santa Lucia. There was a family in that area, which is an extended family of Arcilla and they recruited a lot of people from that area. And I was just really lucky when I had the chance to work for PPI. Because I saw this big line when I was passing through [Santa Lucia]. I was working as a factory worker. . . . So I asked one guy and he said they are hiring in Iraq for PPI. And I applied." One consequence of this extensive Pampanga-based recruiting network was a remarkable spatial concentration of PPI's Filipino workforce. Prior to the imposition of the travel ban in summer 2004, roughly 70 percent of its workers came from the Pampanga region, according to AES records.[6] The effect could be even more pronounced at the level of a *barangay* containing perhaps a few thousand people, with dozens working on bases in Iraq.

At this point it may be useful to discuss a distinction among military contractors that shapes labor needs and one's recruiting and work experience. In most cases companies either receive a direct contract from the military to provide specific services—for example, PWC, which was tasked with shipping food to bases in Iraq through its massive DLA contract—or are subcontracted by a prime contractor like KBR for certain tasks, such as running DFACs on a base, which is what Serka was subcontracted to perform. Companies like Serka and PWC come to the Philippines looking to fulfill well-defined labor needs (kitchen staff and truck drivers, respectively). Hence Michelle's boot camp to provide her neighbors with a basic set of cooking skills and knowledge that would allow them to pass the screening process devised by Serka's recruiting agency.

A second type of logistics firm is what I call *body shops*. Body shops are companies that have multiple contracts or subcontracts covering a range of responsibilities. Several of KBR's largest subcontractors in Iraq, such as PPI, Kulak, and GCC, began as or evolved into body shops. Another prominent body shop in Iraq was First Kuwaiti General Contracting. In addition to holding several subcontracts with KBR, First Kuwaiti was also a significant DoS contractor whose tasks included construction of a massive new embassy in Baghdad and running the embassy's DFAC for the security guard force. In contrast to companies like Serka that have specific labor requirements, body shops provide a variety of services requiring a large pool of unskilled or semiskilled labor.[7] A partial accounting of work performed by PPI employees I interviewed is illustrative. Their jobs have included construction, washing laundry, serving food in DFACs, guarding Iraqi day laborers, cleaning soldiers' living quarters, running MWR facilities, cleaning latrines, driving buses on bases, and cataloguing inventory in warehouses.

Filipinos' accounts of the recruiting process and life on bases make clear that body shops tend to see their workers as fungible commodities that can be deployed and redeployed to perform whatever task has the greatest immediate need. Neither Sam nor Carlos, for instance, worked in the fields that they originally applied for. In Sam's case he was told by the local recruiter in Santa Lucia that PPI was looking for administrative assistants. When he got to the main office in Manila, AES staff said they wanted masons and carpenters. Because of his experience in a factory the company decided to hire him as a construction "engineer." Another early PPI hire, Angel, was recruited as a warehouseman. After a year of this work, he recalls, "PPI needed [LN] escorts, so they trained us." This job entailed going outside the base and picking construction day laborers from among the throngs of Iraqis looking for work: "We would go outside the gate. And we were escorted by military because we don't have a gun. And we go outside and if the company need 100 person we select there. We went to a place like a cottage where there was more than a hundred people sitting around waiting. And we would pick the workers needed." Angel was chosen for this job because he spoke some Arabic as a result of time working in Saudi Arabia in the late 1970s and doing reconstruction and cleanup work "in Kuwait in 1991 after the war."

PPI workers hired during those early hectic months also describe a minimalist recruiting and vetting process in which an applicant's skills and experience were secondary considerations. Flora recalls that interviews—if one can even call them that—were conducted in groups of ten to twelve applicants at once. The first involved an AES employee who "asked how we found out about the job. And that's it." Following this she was escorted with the other applicants into a room with Neil Helliwell. "The only question he asked was, 'Aren't you afraid of going to a warzone?' I said, 'No.'" Shortly afterward she was informed that she was hired. Another PPI employee, Fidel, also remembers meeting Helliwell with ten other applicants: "The interview was not very hard. Just, 'OK. You want to go to Iraq. Why? Are you willing?' 'Yes sir, I'm willing sir.' 'OK. What's your category? What do you know? What's your job?' Like this. . . . At that time if any position is available you grabbed it. It was very easy because this Neil Helliwell, he knows that Filipinos are—what did he call it?—we can be put in a different place, very easy to train, like, 'flexible Filipinos.'" Fidel applied as a warehouseman and forklift driver, which matched his experience. Upon arriving in Iraq he was assigned to housekeeping, where he worked for the next two years before transferring to a warehouse position.

That body shops figure prominently in cases of trafficking and egregious labor abuses in Iraq and Afghanistan is not unrelated to their dehumanizing view of workers as commodities. First Kuwaiti, for example, recruited dozens of Filipinos to work on its embassy project in Iraq under false pretenses, originally promising them jobs at luxury hotels in Kuwait and Dubai. As for PPI, according to one former KBR administrator, in 2004 its man camp at Victory Base Complex in Baghdad "looked like a concentration camp" with workers standing in lines waiting to be served "curry and fish heads from big old pots" and eating "outside in 140 degree heat."[8]

Despite this, many I talked with echo Carlos and Fidel in describing themselves as "lucky" to get hired by subcontractors like PPI and Serka. They cite several factors that makes employment on military bases in the region more desirable than similar positions for civilian projects in Middle East. Perhaps the most important is that these jobs tend to pay more than nonmilitary work. A salary of $600 a month for washing laundry, $450 working as a kitchen assistant, or $800 as a baker could be $100–400 more than the same job in Saudi Arabia or the UAE. Moreover, companies in those countries often deduct expenses for either accommodation or food, resulting in an effective monthly wage $100–200 less than stated in a formal contract. Such deductions are not applied to workers on military bases. Similarly, for positions in the Middle East, recruiting agencies often charge successful applicants fees of several hundred dollars. These fees ostensibly cover processing and labor expenditures. They also bolster profit margins, with foreign employers offloading these costs onto labor migrants. In 2003–4 military contractors were so desperate to quickly amass large workforces that they instructed their agencies in the Philippines to waive recruiting fees, instead paying them sufficient amounts per employee to cover both processing costs and profit margins. In fact, Serka's original agency partner, Blazing Star, was fired when the company learned that it was still charging recruits several hundred dollars despite these extra payments.

A final factor is the relaxation of age restrictions. Several people with experience in the region told me that companies in the Middle East refuse to hire older workers, with cutoffs ranging between thirty-five and forty years of age depending on the company and industry. On bases in Iraq and Afghanistan these restrictions have tended to be substantially looser. One Serka worker I interviewed was nearly sixty when he was hired. Another PPI employee—who worked in the region for more than a decade prior to 2003, including for military contractors in Kuwait repairing oil infrastructure following the first Gulf War in 1991—recalls, "At the time I was forty-nine, and Anglo [AES] had an age limit of fifty, because it was a warzone. At that age, in other places, work is not allowed."

These factors—especially the lack of recruiting fees and relaxed age restrictions—reflect the immense pressure subcontractors were under to assemble a large pool of workers to perform contracted tasks in 2003–4. As noted in chapter 2, KBR's bid for the LOGCAP III contract stated that the company would self-perform most of the required work. While this was feasible for relatively smaller operations like the peacekeeping missions in Bosnia and Kosovo, such plans were overwhelmed by the immensity of labor requirements to support military campaigns in the Middle East. Following the end of the first phase of operations in Iraq in May 2003, the military insisted on a rapid scaling up of logistical support. That month Army officers ordered KBR to establish more than thirty DFACs at bases across the country, with the expectation that troops would be able to eat "franks and beans" and other hot food by July 4th. In the ensuing months the company "went from supporting tens of thousands to supporting hundreds of thousands," necessitating the turn to subcontractors.[9]

Another problem KBR and other military contractors faced concerning labor was that they were initially barred by the U.S. government from hiring Iraqis due to security concerns. Indeed, PWC's first response to the travel bans in 2004 was to inquire with the U.S. embassy in Kuwait about the possibility of hiring Iraqi drivers, a request that was turned down. Former KBR supervisor Mike Lamb also identified this as a key reason his company turned to subcontractors from the region to acquire the necessary labor: "Going into this war, the original intention was to use Iraqis for labor. . . . We were going to use locals. We were going to help the economy. We were going to hire people who were unemployed. But for security reasons we hired labor that has been working in the Middle East for decades. Indonesians, Filipinos, Indians . . . had been working in countries like Kuwait, Dubai, Saudi Arabia. And they were [working for] companies that already had the supply lines for the labor."[10] With a supportive government and a well-developed recruiting industry, the Philippines was a logical supply line to turn to for labor needs.

No company assembled a larger workforce in these first few hectic months than PPI. By early September 2003 its recruiting efforts in the Philippines were in full swing. In the ensuing seven months the company sent more than 4,000 workers to Iraq, an average of nearly 150 a week. Everyone I talked with who was hired by the company in this period highlighted the crowd of applicants outside the AES office. According to Nicky Arcilla, at its peak the agency was receiving 1,500–2,000 applications a day.[11] Thousands of people like Danilo—mostly from Pampanga and towns near Manila—slept on the street and sidewalks outside the agency because it was too difficult to go home every night and they feared losing a job if their name was called when they were not there. Eventually nearby residences started to sell food and offer use of their showers and toilets for a fee.

It was not just the recruiting process that was rushed. Isko recalls an equally fast deployment schedule: "After your name was called you would be sent directly to the rooftop [of the AES office] and not allowed to go home again. Because tomorrow might be a flight. Once the flight was scheduled you went to the airport and signed the contract there. After this we went directly to the special immigration lane for PPI workers. All passengers on the plane were PPI—200 plus!" Such expediency was facilitated by foreign embassies and the Philippine government, who worked closely with recruiting agencies and subcontracting companies to speed up processing. Though not going into detail, Arcilla acknowledges that his company received "special privilege[s] . . . to process them [workers], expedite the papers" from the POEA.[12] Likewise, at the peak of Serka's hiring binge the Turkish embassy devoted resources to process more than a 100 visa applications in less than a day, a task that would typically take a week.

When the Philippines imposed its travel ban to Iraq in August 2004, authorities estimated that nearly 5,000 Filipinos recruited through official channels were already working in the country for military contractors (this total did not include truck drivers hauling goods from Kuwait to Iraq), with 1,000 more in transit at Dubai and another 6,000 "in the pipeline to go to Iraq."[13] Eventually a good number of these 7,000 people caught up in the ban would find their way to bases in Iraq or Afghanistan, either through their own means or with the help of recruiters that continued to ply their trade for military contractors despite the ban and threat of delicensing. That, though, is a story to be told in chapter 7. Before this I need to describe the Bosnian recruiting process, which differs significantly from the Philippine one.

Navigating Websites, Hotels, and Shifting Prime Contractors

Whereas the Philippines has a robust recruiting ecosystem and a well-developed institutional framework that facilitates labor export, companies looking to the Balkans for labor have had to develop their own procedures and provide their own recruiting manpower. Consequently there are significant differences in the recruiting process in the region, as well as the type of military contractors that seek labor. These differences are manifest in the distinctive narrative anchors and spaces highlighted by those I talked with in Bosnia. In contrast to Filipino workers, whose world is populated by subcontractors, recruiting agencies, and government policies and bureaucrats, Bosnian narratives stress a shifting constellation of LOGCAP prime contractors, websites, and hotels scattered across multiple continents.

In comparing experiences of Bosnian and Filipino laborers, it is necessary to start with the observation that the Bosnian state is completely absent in either regulating or facilitating overseas labor recruitment by foreign firms. This stark contrast with the Philippines has a number of implications, not least concerning how one can periodize recruiting practices and histories for these two flows of labor. For Filipinos the key disjuncture centers on the Philippines' imposition of travel bans—first to Iraq in 2004 and subsequently to Afghanistan in 2007. Those hired prior to the bans encountered a recruiting process and actors that are broadly similar to those who apply for similar jobs in Asia and the Middle East, with the exceptions highlighted above. Following the bans, recruiting was pushed underground, altering pathways and increasing risks for both workers and local recruiters. In Bosnia the critical juncture is the transition from the LOGCAP III contract to its successor, LOGCAP IV, at the end of 2008. This transition, as I described in chapter 4, led to a significant downshift in pay and status for Bosnians recruited under the new contract, particularly those who work for DynCorp in Afghanistan. It also altered the recruiting and deployment process, which has become more truncated.

Under the LOGCAP III contract, KBR, which was the sole prime contractor, was the main recruiter of Bosnian labor.[14] This was directly related to Brown & Root's logistics support for U.S. peacekeeping forces in Bosnia since 1996. Indeed, most of those recruited during the first years of the wars in Afghanistan and Iraq transitioned directly from jobs with the company in Bosnia. This was fortunate timing for those who made the jump, as the peacebuilding mission was beginning to wind down in the early 2000s. Moreover, Bosnians who joined KBR's projects in CENTCOM experienced a substantial uplift in pay and status, with salaries that were comparable to American employees holding similar job titles.

All Bosnian KBR hires under LOGCAP III—whether existing employees or new hires—were required to go through a lengthy recruiting and deployment process. The first step was submitting application materials online at the company's jobs site. This was necessary even for those who found out about an opportunity from friends or former managers who had transferred to Iraq or Afghanistan. In fact, during the mad dash to acquire labor in 2003 KBR temporarily placed a moratorium on hiring workers who were still employed by the company in Bosnia due to the large number who were being poached to join projects in the Middle East. After posting a résumé, applicants would wait for an email or call from KBR recruiters asking them to come to a hotel in Sarajevo for interviews. Those who passed the interviews and received a job offer then waited for KBR to arrange a U.S. visa and flight to Houston where they underwent medical tests and received several weeks of training at KBR's Greenspoint Mall deployment center alongside U.S. recruits. Once this was complete new hires were flown to Iraq, Kuwait, or Afghanistan.

Greenspoint Mall and its surrounding environs has a reputation for being run-down and violent, having suffered the fate of many other suburban malls in the U.S. in recent decades. One news story about KBR's American employees put it this way: "Dimly lit and often eerily vacant, Greenspoint isn't an ideal place to spend one's last weeks before going off to war. The mall can't shake its old nickname—'Gunspoint'—it took on after a spate of violent crimes in the mid-1990s. A few days after Thanksgiving 2007, as the holiday shopping season began, Greenspoint was evacuated after a murder-suicide at the Body Luxuries lingerie store."[15] Several Bosnian KBR employees had similar impressions of the area. Elvis sarcastically recalls: "It was such a safe area [the mall] that they actually had to put a police station in it. That weekend as I arrived they shot the cop. . . . The place is eerie. Between six in the morning and four in the afternoon there is not a soul alive." Another individual recounted an attempted mugging as he walked from a nearby hotel to KBR's deployment facility in the shuttered Montgomery Ward department store. Despite this, others enjoyed their time in Houston. Fedja remembers that the large number of recruits from Tuzla made the deployment process feel "like on a school camp . . . 70 per cent of the people I knew there." For Sanja, the deployment center is where she met Laura, a KBR recruit from Houston, "one of the best people I have ever met in my life. I can call her my best friend."

As KBR expanded its recruiting in Bosnia during the 2000s, more and more of those it hired had no prior experience working with the company or other military contractors. This situation was even more common with the second wave of hires stimulated by the LOGCAP IV contract, which awarded Fluor and DynCorp support responsibilities for Afghanistan. One consequence of this shift is that websites and online forums have become increasingly important exchanges for information, whether rumors about employment opportunities, discussions of working conditions with different companies, or suggestions for navigating the recruiting process. The most popular of these websites, slobodni.net, hosts a dedicated, moderated forum titled "LOGCAP Poslovi" (LOGCAP Jobs). Threads and posts within this forum range widely. One can find a copy of DynCorp's test to determine English-language proficiency (along with an answer key); information about technical exams for those applying for electrician or plumbing positions; rates and qualifications for *sudski tumači* (court interpreters) in Tuzla who can provide official translations of police reports required for background checks; updates on pay scales for specific positions and projects; discussions of the conditions that cause one to fail health exams (high blood pressure and bad teeth are the most common culprits); memorials for compatriots who have died while working in Iraq and Afghanistan; and detailed debates about company policies, such DynCorp's decision to offer Asian contracts to Bosnian applicants. This last topic is the subject of a separate thread that has generated nearly 100 posts, which have been viewed more than 12,000 times. These numbers are dwarfed by the general threads for KBR, Flour, and DynCorp, which combined have more than 22,000 posts that were viewed more than 3 million times between late 2009, when LOGCAP Poslovi was established, and July 2017.

Web portals that cover local news are other key sites for information. A Lukavac-based portal, sodalive.ba, for instance, has published dozens of articles on recruiting events in Tuzla, life on military bases as a contractor, and the effect that this phenomenon has had upon economic and social relations in region over the past decade, with headlines such as "Recruiters for Fluor Have Arrived in Tuzla" (February 6, 2017), "Lukavac Residents' Search for a Better Life Leads to Afghanistan" (January 31, 2012), and "Youth in Bosnia and Herzegovina— Afghanistan or a Luxury Cruise Ship?" (March 11, 2013).[16] This last article highlights the disillusionment of youth in Bosnia, who are increasingly desperate to leave the country due to high unemployment, low pay and the political situation, comparing the experiences of those who choose to work as waiters or hospitality staff on cruise ships and those who sign on with Fluor or DynCorp in Afghanistan. The article struck a nerve among readers, generating nearly fifty comments,

most of them about work and life in Afghanistan. For several years sodalive.ba
was edited by a former KBR employee from Lukavac who worked for more than
a decade with the company starting in 1996. In 2016 he again left Lukavac to work
for a U.S. PMC that has several logistics contracts to support operations against
ISIS in Iraq and Syria.

In addition to websites, hotels loom large as significant spaces in nearly every
account. From the Holiday Inn at Sarajevo to the Marriott and the Wyndham in
Greenspoint, and from the Mövenpick and the Grand in Dubai to the Le Meridien
in Tashkent, Uzbekistan, Bosnians have circulated through different networks
of hotels which define, in part, distinct recruiting processes and deployment pathways developed by each military contracting firm. Indeed, by the end of my interviews in Bosnia I found that I could reliably identify the company people worked
for, as well as period of employment, just by the list of hotels—and activities that
took place at them—that they mentioned in their stories. Hotels, in other words,
have constituted critical infrastructural nodes in the accumulation of military
labor from the Balkans, giving them a geopolitical and geo-economic significance that has not been adequately appreciated to date.[17]

Hotels serve a variety of functions. First, and most significantly, all three LOG-
CAP prime contractors (KBR, Fluor, and DynCorp) use them as bases for recruiting in the Balkans. Rather than establish permanent offices in the region, the
companies rent out blocks of hotel rooms or conference spaces that serve as
temporary recruiting centers. KBR alternated between two hotels in Sarajevo:
the Hollywood, located next to the airport and the headquarters of the NATO-
led peacekeeping mission, and the downtown Holiday Inn, made famous as the
home of international media covering the siege of the city in the early 1990s.[18]
Fluor and DynCorp use Hotel Tuzla in downtown Tuzla, which has the advantage of putting them at the epicenter of the country's military labor pool.

Hotels also operate as sites for predeployment training and as waystations while
waiting for necessary paperwork. One KBR employee, Rena, recalls a cohort of
recruits before hers waiting for transit visas to Dubai for three months at the Marriott in Houston. The cause of this delay was temporary travel restrictions the
UAE put upon people from the Balkans following a spectacular jewelry heist by
the infamous Balkans-based "Pink Panthers" gang in 2007.[19] To save money and
speed up the deployment process to Afghanistan, both Fluor and DynCorp have
rejected KBR's strategy of sending workers to the U.S., instead flying them to
Dubai for health exams and abbreviated training courses held at hotels like the
Mövenpick or Grand (DynCorp), or bringing in staff to conduct training at the
Hotel Tuzla before deployment (Fluor). In addition to saving money, this indicates the lower status that Bosnian hires have experienced while working for the
two companies, who unlike KBR have drawn a clear line dividing American from

non-American direct hires. For its Ebola support mission in West Africa, and more recent contracts to provide logistics support for SOF forces and drone operations in the Sahel and Horn of Africa, Fluor has moved its training program to hotels in Dubai, where workers also wait for necessary visas before flying to Niger, Uganda, Cameroon, and Somalia.

Finally, hotels figure prominently as rest and recreation sites. As part of its benefits package, KBR offered employees in CENTCOM three paid vacations a year under its LOGCAP III contract, including covering travel costs between military bases and home. This policy has been followed by Fluor and DynCorp—though in reduced form by the latter which offers two vacations a year, with travel expenses covered only for the first. For those working in difficult warzone conditions, a night or two layover in Dubai, Istanbul, or Tashkent presents an opportunity for shopping, entertainment, or just lounging at the hotel's pool. Shopping is especially popular for those routed through Dubai, who load up on goods at the malls to bring back as gifts for family and friends. Elvis, who worked in Afghanistan in the early 2000s, remembers KBR using a two hotel system in Tashkent, which was the main hub for transit to and from the country at the time: "It was the Sheraton and Le Meridian. One was for [people going] in, the other was for [people going] out. They didn't want these two groups of people mixing." According to Elvis, Le Meridian, the outbound hotel, took full advantage of regular flights of workers leaving Afghanistan with money to burn and looking to "blow off steam." "The Le Meridian was probably the most expensive hotel on the face of the planet at the time. They were charging six bucks for a can of Heineken . . . Breakfast was 15 dollars. Hookers were everywhere. Shit, every night there was like a platoon of them. Of course suckers were falling in love, spending their money, drinking their money, and going back to Afghanistan broke, dying." Here the contrast with Filipinos employed by subcontractors is instructive. These companies do not offer vacations—paid or unpaid—forcing workers to stay on bases without trips home for the length of their contract, often two or more years. Additionally, while subcontractors are obligated to pay for flights to and from home, they do not cover accommodation during layovers, forcing workers to sleep in the airports.

As noted above, the transition from LOGCAP III to LOGCAP IV represents a significant dividing line for Bosnian workers, with the Pentagon pushing prime contractors to lower salaries for direct hires from Southeast Europe. Even KBR lowered its pay scale for new Bosnian recruits to Iraq and Kuwait. Lena, who was hired in late 2008, remembers: "They told us in the middle of the processing [in Houston] that they were going to cut our pay. People started to scream and yell." This led the company to delay implementation of this decision until the group after hers. KBR lost Afghanistan to Fluor and DynCorp under the new contract, and the Iraq War was winding down, so its recruiting efforts dropped off substantially at this point. Far more people have been affected by the differential pay policies implemented by Fluor and DynCorp, which make a distinction between American, West European (primarily Britons), East European, and Asian direct hires.

DynCorp's decision to offer both European and Asian contracts to Bosnian workers beginning in 2010 has been especially contentious. One issue, obviously, is the dramatic difference in pay, with Asian contacts paying $12,000 to $18,000 a year and workers performing the same jobs under a European contract typically earning three times this amount. What angers people the most, however, are the false promises made by local recruiters—that is, Bosnians hired by DynCorp to run recruitment in the country, under the supervision of an American manager—that applicants who signed an Asian contract could easily switch over to European positions and pay when they arrived in Afghanistan. In one case a local recruiter was beaten by family members of a worker who found out he had been duped. Several DynCorp employees told me they were convinced that local recruiters' pay was linked to the number of people they could convince to sign Asian contracts, given the amount of money the company could save in labor costs.

The Logistics of Assembling a Transnational Workforce

While the concept of logistics originates with the supply of military operations, it has taken on wider connotations over the past century. As Edna Bonacich and Jake Wilson observe, "Its meaning has been expanded to refer to the management of the entire supply chain, encompassing design and ordering, production, transportation and warehousing, sales, redesign and reordering. This entire cycle of production and distribution is now viewed as a single integrated unit that requires its own specialists for analysis and implementation."[20] Thus from a commercial standpoint logistics is concerned with the circulation of commodities, getting goods—be they sneakers, flat screen TVs, or cars—from one place to another. With the advent of global supply chains this is an incredibly complex and diffuse process, involving a "network of infrastructures, technologies, spaces, workers and violence that makes the circulation of stuff possible."[21] This is what Deborah Cowen means when she refers to the production of commodities today as occurring "*across logistics space* rather than in a singular place."[22]

This focus on the logistics of commodity production and distribution is insightful, but I want to argue for a broader consideration of logistics. One that is attendant to both goods *and* people. For logistics does not just underpin the transnational circulation of goods, but—increasingly—labor as well. Consider the cargo ships that haul goods across the oceans. As the shipping industry globalized in recent decades national carriers have been displaced by companies that register ships under a flag of convenience in countries like Panama and Liberia. This allows them to avoid stricter regulations, especially concerning labor costs and standards, imposed by traditional shipping centers like Britain and Greece. Ship owners also began outsourcing operations to other firms that are responsible for assembling crews that now often resemble a veritable United Nations of labor.[23] As with U.S. military logistics workers, Filipinos constitute one of the largest contingents of seafarers, and the sites, processes, and actors involved in recruiting labor for both industries are remarkably similar, though the POEA does separate out land- and sea-based recruitment and employers for administrative purposes.

Transoceanic shipping, in short, is now highly dependent on the acquisition of a global workforce. It is just one of many economic sectors; others that draw extensively on transnational labor supply chains to staff their workforces include cruise ship operators, transoceanic fishing fleets, logistics firms supporting humanitarian and peacekeeping operations around the world, and large corporations that dominate worldwide oil, gas, and mineral extraction. To these industries can be added the massive labor import-export regime between wealthy Gulf petro-states and poor Asian labor-exporting countries. All together this admittedly partial accounting of the phenomenon represents millions of people circulating through different, but frequently overlapping, transnational labor supply chains.

The point I want to make here is that for these industries and countries, as with military contractors, assembling a global workforce is itself a complex undertaking, one populated by its own logistics spaces and labor, including recruiting agencies, websites, transportation companies, hotels, government bureaucrats, and labor brokers. While there is a great deal of excellent research concerning those who perform logistics labor for commodity production and distribution—such as driving trucks, sorting and packaging goods at warehouses, and unloading cargo at ports—to date the logistical infrastructures and labors involved in assembling large-scale global workforces have not attracted the attention of those who study logistics. This is an oversight, I believe. Certainly, when it comes to the global labor system that the U.S. depends on to maintain its overseas military empire, both the Serka employee who serves food at a DFAC at Al Udeid Air Base in Qatar, and the Manila recruiting agency that processed her, are equally important parts of the story.

DARK ROUTES

Because of you I became very rich. Otherwise, if you did not help me get workers from the Philippines how can I have my business inside the base?

—Senior executive of a logistics subcontracting company to an underground recruiter in Manila

On May 15, 2004, the U.S. embassy in Kuwait convened a briefing with representatives from major contractors providing logistics support for the U.S. military in Iraq and Kuwait. This meeting followed stories earlier that month in the Indian newspaper *Hindustan Times*, alleging trafficking and physical abuse of Indian citizens working at U.S. bases in Iraq. Impressing upon the assembled group that "the scandal was very big in India," embassy officials encouraged firms to their review hiring and work practices and respond to Indian embassy inquiries concerning the status of their employees, noting that "a forward-leaning tack by U.S. contractors" would be "good public relations, at the very least."[1] Days later India imposed the first travel ban to Iraq by a labor-exporting state.

From U.S. diplomats' point of view, the most explosive claim in the news articles was that U.S. troops beat Indian laborers—who worked as cooks at the U.S. base Q-West (also known as Endurance) near Mosul—and facilitated their trafficking into Iraq. The embassy in New Delhi mobilized immediately following the publication of these articles in early May, tracking down the named workers and their recruiting agency and conducting its own interviews. Two days later it composed a widely distributed cable detailing findings. Titled "Mission Debunks Media Reports of Abuse of Indian Workers by the U.S. Army," the document triumphantly declared that it found the reports "to be exaggerated and largely false," concluding that "there is no evidence that American soldiers were part of the trafficking of these workers" and no evidence "they were beaten by Americans."[2] What is more interesting is what this investigation confirmed—specifically, the workers' allegations of trafficking and physical abuse.

The outline of this series of events is as follows. In August 2003 several men each paid a Mumbai-based recruiting agency, Subhash Vijay Company, $1,600 to obtain food preparation jobs in Kuwait. According to the agency, it obtained visitor visas to Kuwait for the men for GCC, which stated that they would be converted to worker visas upon arrival. Instead the Indians were met by unknown men at the airport in Kuwait, their passports and papers were confiscated, and they were loaded into a van that drove them to Q-West. One of the men interviewed by embassy officials described his experience on the base as "unmentionable," while another stated that their supervisors made them work up to twenty hours a day and would beat them if they did not agree to work these hours. Responding to these facts the cable tepidly concluded that the embassy's investigation "indicates these Indians appear to be victims of unscrupulous Indian and Gulf-based manpower agencies" and that "it is possible" U.S. contractors "had overall responsibility for conditions" at Q-West (which the embassy mistakenly identified as "Crew West"). Following this investigative 'exoneration,' DoS and DoD officials turned their attention to cajoling the government of India to rescind its travel ban, efforts that met with temporary success in early June after military officials presented it with a draft of protocols to improve living conditions for TCNs in Iraq.[3]

Most of the egregious examples of trafficking and other labor abuses have been perpetrated by subcontracting firms like GCC that hail from Turkey or the Gulf states. Companies from the latter countries, in particular, are notorious for abysmal treatment of their largely South and Southeast Asian workforces. Human rights organizations such as Amnesty International and Human Rights Watch have repeatedly documented labor abuses suffered by foreign workers in the region, with the problem particularly widespread in the booming construction industry.[4] In relying on firms like GCC to provide labor, the military has in effect imported these exploitative labor practices onto its bases in Iraq, Afghanistan, and other countries in the region. At the same time, it has continuously tried to minimize responsibility for trafficking and other labor abuses committed by its contractors, instead defining oversight authority and jurisdictional powers in the narrowest possible terms. Consequently, these abuses have continued to be perpetrated by contractors and recruiting agencies in the decade and a half since they first emerged in Iraq. But the problem of abusive labor and recruiting practices need also be understood in relation to the confluence of two further dynamics connected to changes in military contracting in recent decades: 1) the offshoring of labor and 2) the downsourcing of risk. The former is directly related to the shift in logistics labor from uniformed and American to civilian and foreign, while the latter is a product of complex and lengthy labor supply chains associated with subcontracting that foster moral detachment and obfuscate responsibility for labor conditions.

This chapter explores these issues in three parts. First, I describe instances and types of trafficking and labor exploitation perpetrated by military contractors, situating them within the dynamics of offshoring and downsourcing. Next, I examine illegal recruiting practices that flourished following the 2004 travel ban to Iraq and 2007 ban for Afghanistan imposed by the Philippines. These bans pushed recruiting of labor to bases in those countries underground, increasing workers' precarity. Yet the experience of Filipino workers I interviewed also demonstrates a range of agency and initiative on the part of those who traversed these dark routes. Their stories also highlight the role that military contractors have played in undermining labor-exporting countries' travel bans. I end the chapter by returning to the question of oversight, analyzing rationales by military officials to justify minimal oversight responsibility for trafficking by their contractors and drawing parallels to the offshoring of labor and supply chains by U.S. corporations in recent decades.

Offshoring Labor and Downsourcing Risk through Subcontracting

Trafficking of workers to Iraq, Afghanistan, and other military bases in CENTCOM has continued unabated since KBR first decided to outsource most of its labor needs to subcontractors in 2003. More than a decade later, for instance, an investigation by journalists Samuel Black and Anjali Kamat found substantial evidence that employees of Fluor's and DynCorp's subcontractors in Afghanistan were deceived and exploited by recruiting agencies. Of the seventy-five current and former workers they interviewed, "65 said they paid agents fees ranging from $1,000 to $5,000. Many said their monthly salaries, generally $400 to $800, ran several hundred dollars short of what they were promised. Some paid fees, only to be warehoused by an agent for months and never receive a job. Nearly everyone we talked to was still paying back loans."[5] Three years before Black and Kamat published their findings, former military auditor and lawyer Sam McCahon testified before Congress that such practices were widespread, and hence the U.S. government was responsible for allowing trafficking to flourish on its bases in the region.

> Even though he [a victim of these trafficking schemes] now knows he
> was deceived, he is helpless. If he speaks to anyone with the government
> he is terminated immediately and sent home. (The prime contractor

typically instructs its employees that they are forbidden to inquire or report trafficking conditions of subcontractors, thereby completing the conspiracy of silence and mitigating detection of the crime.) The victim cannot quit because he has the outstanding loan to the loan shark. He must remain, working 12 hour days, 6 to 7 days per week in the combat zone. By the time he completes two to three years, he has still not retired the debt. He is an indentured servant to the U.S. government contractor.[6]

As McCahon noted, these debts place workers in a position of involuntary servitude—unable to refuse jobs they are given regardless of working conditions, deception about place and type of employment, or discrepancies between promised and actual salaries. Another way to describe this phenomenon is debt bondage, a form of trafficking to which "workers around the world fall victim . . . when traffickers or recruiters unlawfully exploit an initial debt the worker assumed as part of the terms of employment." Significantly, the definition I quote here comes from DoS's annual *Trafficking in Persons Report*, which notes that such practices are illegal under U.S. law.[7]

One of the more notorious cases of trafficking, which is illustrative of the scale and depth of the problem, occurred in 2008 in Iraq. That spring a Kuwaiti subcontractor, Najlaa International Catering Services, received several DFAC contracts from KBR. In anticipation of labor needs, it contracted with manpower firms in India, Sri Lanka, Nepal, and Bangladesh to recruit approximately 1,000 workers who paid up to $5,000 each in exchange for the promise of work in Iraq. Najlaa flew the men to Baghdad, confiscated their passports, and put them up in a windowless warehouse, where they spent three months—without pay—due to unrelated mobilization failures that caused Najlaa to be unable to begin its DFAC contracts in September as planned. In December the men staged protests outside of the warehouse, which was adjacent to Baghdad's international airport, bringing their situation to the attention of reporters. The next month KBR rescinded its contracts with Najlaa, leaving the men without jobs. Eventually, under pressure from the U.S. government, KBR found work for several hundred of them and arranged for the repatriation of the rest. However, for those who were repatriated the ordeal was far from over as they still owed thousands of dollars to creditors with little chance of earning enough to repay their loans.[8]

Body shops figure prominently in cases of trafficking and labor abuses. In 2006 First Kuwaiti recruited dozens of Filipinos to work on its embassy project in Iraq under false pretenses, originally promising them jobs at luxury hotels in Kuwait and Dubai. Rory Mayberry, a former company medic, testified about this scheme before Congress in 2007.

First Kuwaiti managers asked me to escort 51 Filipino nationals and to make sure that they got on the same flight as I was headed to Baghdad. Many of these Filipinos did not speak any English. I wanted to help them to make sure that they got on the flight ok, just as my managers had asked me. We were all employees of the same company was my feeling. But when we got to the Kuwait Airport, I noticed that all their tickets said that we were going to Dubai. I asked why. A First Kuwaiti manager told me that Filipino passports do not allow Filipinos to fly to Iraq. They must be marked going to Dubai. The First Kuwaiti manager added that I should not tell any of the Filipinos that they were being taken to Baghdad.

As I found later, these men thought that they had signed up for jobs to work in Dubai hotels. One fellow I met told me in broken English that he was excited to start a new job as a telephone repairman. They had no idea that they were being sent to do construction work at the Embassy.

Well, Mr. Chairman, when the airplane took off and the captain announced that we were headed to Baghdad, all you know what broke out on the airplane. The men started shouting. It wasn't until the security guy working for First Kuwaiti waved an MP-5 in the air that the men settled down. They realized that they had no other choice but to go to Baghdad.[9]

Mayberry's testimony was echoed by John Owens, a First Kuwaiti construction manager on the project, who recalled boarding a chartered flight from Kuwait to Baghdad and noticing that all of the other company workers had tickets that stated that their destination was Dubai. When he asked another manager about this he was told, "Don't say anything. If Kuwaiti customs knows they're going to Iraq they won't let them on the plane."[10] When the plane landed in Baghdad all of the men were smuggled past customs into the Green Zone.

In addition to trafficking, workers in Iraq and Afghanistan have frequently experienced a host of other labor abuses ranging from squalid and inadequate living conditions to wage theft. Owens observed First Kuwaiti workers "verbally and physically abused" and having "their salary docked for as much as three day's pay for reasons such as being five minutes late."[11] Wage theft is perhaps the most common form of exploitation perpetuated by subcontractors. This takes two primary forms. The first involves paying wages that are substantially less than initially promised. The journalist Sarah Stillman provides a good example of this in her gripping account of two Fijian women, Vinnie and Lydia, recruited to work in Dubai in 2007 with promises of salaries ranging from $1,500 to $3,800 a month.

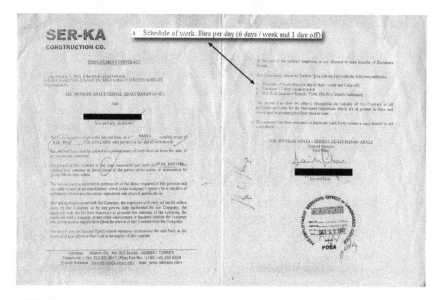

FIGURE 7.1. 2003 Serka contract for Filipino worker in Iraq

When they arrived in Dubai the subcontractor that they worked for, Kulak (a body shop from Turkey), told them the jobs were actually located in Iraq and they would be paid only $700 a month. After being passed on to another Turkish subcontractor their pay was again cut in half.[12] Another variation of this form is reneging on promised salary bumps after a promotion or a designated length of time working for a company.

A second form of wage theft involves refusing to pay for overtime or hours worked beyond the terms of a contract. For example, contracts that Filipino DFAC workers signed with Serka in 2003 indicated that their schedule of work would be eight hours a day, six days a week, with their monthly salary based upon this forty-eight-hour workweek (figure 7.1).[13] Yet every one of the former Serka workers that I interviewed stated that they worked twelve-hour shifts, seven days a week, with no additional pay for the extra thirty-six hours worked each week. This is not unusual as twelve-hour workdays with no days off is standard for TCNs working for subcontracting companies.[14]

As mentioned above, the provenance of subcontracting companies that have been implicated in the most egregious cases of trafficking is not immaterial. Nearly all hail from Turkey or Gulf states where such practices are widespread. Also relevant, in my view, is that body shops like GCC, First Kuwaiti, PPI, and Kulak view their largely unskilled workforce as a disposable commodity. This is well illustrated by the not uncommon phenomenon of passing workers on to other firms—as happened to Vinnie and Lydia—and Najlaa's abandoning its workers at the warehouse in Baghdad after its contracts with KBR were first delayed and then fell through.

There are two other dynamics that are important for understanding the prevalence of labor abuses in connection to military contracting. The first is what Darryl Li calls the offshoring of military work: "The so-called privatization revolution has also been an offshoring revolution, with U.S. contractors frequently overseeing an even larger set of foreign subcontractors and workers. . . . TCNs in particular work on U.S. bases under military authority while lacking most of the protections of American law, local regimes, or their home governments. They are often employed by non-U.S. companies subcontracted by American corporations, paid a fraction of what American contractors and soldiers make, and can be easily deported if deemed noncompliant."[15] In framing military contracting as offshoring, Li draws a parallel to the strategy of offshoring manufacturing that numerous U.S. corporations have pursued in recent decades. Though there are differences, this is an instructive comparison. The primary driver of offshoring is a desire by companies to lower costs, shifting manufacturing from expensive U.S. labor to cheaper countries like Vietnam, China, and Bangladesh. Savings come not just from lower wages, but also less stringent labor and environmental standards in these countries. The result is that offshore workers are paid significantly less and have access to few of the legal safeguards that U.S. workers enjoy. In similar fashion, military contracting is driving a shift in the composition of the workforce from uniformed and American to civilian and foreign, with an attendant decrease in wages and labor standards. This is especially the case for TCNs, who are caught in a legal limbo in which neither U.S. nor local labor laws apply. In effect, overseas military bases operate as exceptional spaces, jurisdictional voids where these workers have little to no legal recourse when subject to labor abuses by employers.

Another, related, manner in which the lens of offshoring is productive involves the military's desire to disentangle as much as possible logistics operations from legal and political constraints. Here the relevant analogy is the offshore oil and gas industry. As Hannah Appel compellingly demonstrates through her ethnographic examination of oil extraction in Equatorial Guinea, the "offshore" does not merely refer to the location of the oil rigs off the country's coast, but is actively created through various practices that serve to disentangle operations as much as possible from the messiness of the "onshore." The key to this, Appel argues, is modularity—of labor regimes, contracts, technology, and infrastructure—that tends towards "internal containment." She explains:

> Modular or prefabricated structures do not require changing the zoning code but, instead, come with an anticipatory relationship to

place and time—legally compliant, mobile, without foundation, impermanent, and disposable or reusable elsewhere. So too with offshore oil platforms, contracts and subcontracts, and mobile labor forces. These are work-intensive efforts to create juridical and even geographic spaces in which companies can abide by their own rules, bring their own technologies, infrastructures, evidentiary and legal regimes, and people—laborers, lawyers, technicians, consulting firms, specialists, and managers.[16]

As with military logistics, a fundamental aspect of modularity in the offshore drilling industry involves the extensive use of subcontracting and foreign labor, which serves to insulate companies from regulatory oversight.

The second dynamic—the downsourcing of risk—is directly related to complex and lengthy labor supply chains. As the journalist Cam Simpson, who wrote a powerful book detailing the experiences of the twelve trafficked Nepalese men and their families, puts it:

> There is no single villain pulling strings from the top, but instead, several individual actors who make up an overall chain of conduct. It is an inherently transnational enterprise that utilizes a global supply chain extending across multiple countries, and it requires an extensive transnational network of recruiters, contractors, subcontractors, parent corporations, and subsidiaries crossing jurisdictions, countries, and continents. The sheer number of actors involved allows each to point a finger somewhere else—to someone below him in the supply chain, or someone above—or simply to deny his own individual piece of responsibility.[17]

Consider the example of the Indian workers at Q-West that I began this chapter with. Subhash Vijay had hired them to work for GCC, which was a subcontractor for Alargan Group (a Kuwaiti firm), which in turn was a subcontractor for The Event Source (a U.S. firm), which in turn was the company originally subcontracted by KBR, the military's prime contractor (table 7.1).[18] There were, in other words, four layers of subcontractors and recruiting agencies between the workers and the military's prime contractor. The labor supply chain that the two Fijian women, Lydia and Vinnie, traversed was of similar length and complexity. They were first approached by a local recruiter in their neighborhood in Suva, who directed them to Meridian Services Agency, which was recruiting workers for Kulak. Kulak in turned passed the women on to a fellow Turkish firm, Nasa, which was a contractor for AAFES, the DoD organization that manages the military's PX stores around the world (table 7.1).

TABLE 7.1 Transnational military labor supply chains

India Labor Supply Chain	Local recruiter (India)▶ Subhash Vijay (India)▶ GCC (Kuwait)▶ Alargan Group (Kuwait)▶ The Event Source (U.S.)▶ KBR (U.S.)▶ U.S. military
Fiji Labor Supply Chain	Local recruiter (Fiji)▶ Meridian Services Agency (Fiji)▶ Kulak (Turkey)▶ Nasa (Turkey)▶ AAFES (DoD Agency)
Nepal Labor Supply Chain	Local recruiter (Nepal)▶ Moon Light Consultant (Nepal)▶ Morning Star for Recruitment and Manpower Supply (Jordan)▶ Bisharat (Jordan)▶ Daoud & Partners (Jordan)▶ KBR (U.S.)▶ U.S. military

Both of these lengthy labor supply chains are relatively straightforward compared to the convoluted case of the Nepalese workers. As with the Indian and Fijian examples, the twelve men's journey began with a local recruiter, who put them in touch with Moon Light Consultant, a recruiting agency in Kathmandu. Moon Light was working with a Jordanian recruiting agency, Morning Star for Recruitment and Manpower Supply, which was promising work at Le Royal, a luxury hotel in Amman. Upon the men's arrival, however, Morning Star immediately passed them on to Bisharat, a shady Jordanian labor broker that supplied workers for KBR's Jordanian LOGCAP subcontractor, Daoud & Partners (table 7.1). Bisharat housed the men in compounds in Amman for several weeks before arranging a convoy of taxis to take them to U.S. bases in Iraq, a risky journey that the men did not survive.

Following Amanda Wise, I argue that these chains of labor recruiters and contractors—which she points out are characteristic of transnational labor "pyramid subcontracting"—obfuscates responsibility for working conditions and fosters a sense of moral detachment whereby military officials and prime contractors deem labor abuse a problem that largely lies outside their remit. It also contributes to the dehumanizing treatment of workers as a "disposable army" of labor.[19] In short, lengthy labor supply chains engendered by subcontracting induces a downsourcing of responsibility and risk—which ultimately falls on workers themselves in the absence of effective oversight of subcontractors' actions.[20]

Legally, one means through which risk and moral responsibility is downsourced is by reference to "privity of contract," a doctrine that limits the rights or obligations of third parties to contracts. The Army's manual on operational contracting states that when the U.S. military enters a contract with a prime contractor such as KBR, and the prime contractor in turn makes separate contracts with subcontractors,

> The prime contractor has privity with their first-tier subcontractor, but the government has no privity with any of the subcontractors at any tier;

therefore, the government contracting officer cannot direct the prime's first-tier, nor any lower tier, subcontractors. This term is important to the Service commander in that only the prime contractor has direct responsibility to the government. This fact can limit the directive ability of Service commanders, through the cognizant contracting officer, to directly enforce contractor management policies on subcontractors and their employees.[21]

Read cynically, the Army's invocation of privity of contract is very useful for washing its hands of oversight responsibility. And in fact, this is exactly the argument that prime contractors in turn have used to justify their lack of responsibility for monitoring labor abuses by subcontractors. The vice president in charge of contracting for one LOGCAP prime contractor, for example, bluntly told Sam McCahon that his company was taking no measures to mitigate trafficking because "we have no privity of contract with the subcontractors' employees, so it's not our problem."[22] We will return to the question of privity of contract in the final section of this chapter, following an examination of illegal recruiting that flourished in the Philippines after the country's imposition of travel bans to Iraq and Afghanistan.

Underground Recruiting and Navigating the Travel Bans

Illegal recruiting of labor for the U.S. military is not just a matter of trafficking. Indeed, arguably the most traveled dark route has involved the underground recruitment of workers from labor-exporting states like India, Nepal, and the Philippines in the years after they imposed travel bans to Iraq and Afghanistan. Despite the bans, the military directly supported the continued importation of workers from these countries due to its immense labor needs. This is demonstrated most clearly in its efforts to pressure Kuwait not to enforce the travel bans at its border crossings in 2004, as discussed in the introduction. For years the military ignored the overwhelming presence of Filipinos, Indians, and Nepalese working on U.S. bases. Finally, in summer 2010 CENTCOM issued a memorandum ordering contractors in Iraq to comply with "TCN laws" by repatriating workers from countries with existing travel bans (at the time this included Nepal and the Philippines).[23] By then the drawdown of troops was in full swing and thus the need for labor had abated, rendering this shift in stance rather hollow. Later in the year the military extended this order to Afghanistan. In 2011 the Philippines responded by modifying its travel ban to Afghanistan, exempting citizens who had existing contracts with companies working on military bases.[24] Effectively this grandfathered in those already working in the country, while continuing the ban on recruitment of new hires. Yet the flow of Filipino labor to Afghanistan continued.

So how have Filipinos looking for military work evaded Philippine authorities seeking to enforce the travel bans? For many the path still goes through local recruiters and recruiting agencies. Because agencies in the Philippines who recruit for military contractors can have their licenses revoked if they are caught, recruiting has gone underground. To understand how this works I conducted interviews with both workers recruited after the bans were implemented and agencies that have provided labor for three different subcontractors in Iraq and Afghanistan during this period. In some ways underground recruiting is similar to what took place before the bans. Agencies still work with contractors to determine labor needs. They also continue to vet applicants for relevant skills, arrange necessary medical exams, and organize transportation to the region.

Beyond this veneer of business as usual, however, the risks and costs—for both recruiting agencies and workers—can be substantial. One agency I talked with stated that the company they worked for continued to pay them a fee for each worker they successfully deployed, but also allowed the agency to charge applicants a separate processing fee, which amounted to roughly $200 per person. According to a 2006 investigative report by the Army, following the travel ban Serka began deducting $400 from Filipinos' pay after arriving in Iraq, ostensibly to cover travel costs even though the company was already compensated for these expenses in its contract with KBR.[25] These fees and deductions appear to be on the low end according to workers I interviewed, who recall paying up to $3,000 for jobs in Iraq or Afghanistan. Such costs are much higher than the Philippines allows for legal recruiting, where laws stipulate that fees cannot exceed one month's salary. Without question this indicates that supplying labor during bans can be a lucrative proposition, especially for the handful of recruiting agencies and politically connected individuals willing to take the risks. More than one person I talked with, for instance, suggested that President Arroyo's son, Mikey Arroyo, worked with PPI in the early years after the ban to ensure that its workers destined for Iraq would not be detained by officials at Ninoy Aquino International Airport in Manila.

Ensuring the safe passage of workers during bans is an expensive proposition for military contractors and recruiting agencies without such political connections. The head of one agency, whom I will call Edward, told me that "the most difficult one [stage of the underground recruiting process] that we are doing is the airport . . . immigration people really cost a lot of money." For each batch of eight to twelve recruits that he sends he has to pay officials at the airport roughly

$500 to ensure that they are not interdicted. Edward counts himself as fortunate that none of his recruits have been stopped, attributing this to the fact that his contact at the airport is an "honest person" who shares the bribes with coworkers. The consequences for recruiters like Edward if their people are detained at the airport for violating the travel bans are significant. He would lose not only his placement fee from the contractor, but also face the risk of losing his accreditation if someone told authorities who was behind the scheme.

While recruiting agencies have continued to play a role in facilitating labor acquisition after the travel bans, my research suggests that the majority of those who have found jobs during this period have done so without the help of agencies. More commonly, companies with Filipino workers in Iraq or Afghanistan have asked them to spread the word to families and friends about job opportunities when they call home. John, who is from a village in Pampanga, explained to me how one of his cousins arranged work for him and nine other family members this way:

> ME: How did you find jobs in Afghanistan?
> JOHN: Before we went we had a contract. . . . Our cousin there sent us papers to sign. Our cousin went to Afghanistan and found employers for all of us. . . . As soon as she found employers it took a couple of weeks for the papers and then I was gone.
> ME: How did she find these jobs?
> JOHN: She had American friends, and would hear [about opportunities] from other Filipinos at mealtime.

In other instances those who have completed a contract with a firm, such as PPI, are contacted by former managers asking if they want to work again. Alternatively, former employees reach out to companies asking about work. One longtime PPI worker, Fidel, described this last scenario to me: "I sent an email [saying], 'I'm a employee of PPI before in Iraq, so I'm looking if there's a vacancy for equipment operator.' I'm lucky. In 2010, January, they're hiring on Afghanistan. They send me a ticket, [tourist] visa, just in one week, I go fly to Dubai." Rowel, an electrician who began working for PPI in Baghdad in 2004, went home in 2006, and then returned to Iraq later that year, also mentions PPI's use of a tourist visa to get its workers out of the Philippines after the ban was introduced:

> ME: Did you have to buy a tourist visa to Dubai? Is that how you got back?
> ROWEL: I think that PPI, the main office in Dubai [arranged it]. They were telling [officials in UAE] that we got [a] seminar in Dubai. We

got a tourist visa over there so that PPI keeps pushing that, "Oh these guys need to go in Dubai to take a seminar," and then [they] send [us] another country.

ME: PPI was your sponsor for the visa to Dubai?

ROWEL: Yeah.

While PPI provided tourist visas and plane tickets for Fidel and Rowel, they were on their own when it came to clearing immigration at the airport in Manila. Dubai is the primary transportation hub for those illegally working for military contractors, so those with tourist visas to the emirate are given extra scrutiny. If an official believes that someone with a tourist visa is actually travelling to Iraq or Afghanistan they can detain them for questioning or prohibit them from leaving the country. Luckily for Fidel he had a contact at the airport who helped him get through, telling him which line to go to and who to pay to ensure safe passage. When he arrived in Dubai he was greeted by a Filipino who said, "Are you PPI?" It turned out he was not the only one put on the plane by PPI. "There was so many of us. There was a list. I thought I was the only one." Fidel and the other workers were taken to PPI accommodations in Jebel Ali, a port complex in the southwest margins of the emirate. "The first days [were] preparation. You submit your papers, passport, sign contracts, medical [exams]. . . . After medical— so a week [later]—[you are] scheduled to fly to Afghanistan. Every day we just check on the blackboard if there's a flight going to Afghanistan. Every day . . . maybe four people, five people, six people, three people. There's flights going to Kandahar, to Bagram. PPI sent us to different camps."

What's remarkable about Fidel's account is that it illustrates PPI's intimate involvement in facilitating the evasion of the Philippine travel ban, from the procurement of tourist visas under false pretenses to purchase of airline tickets for Dubai. It also demonstrates the scale of the company's underground labor acquisition scheme, with the daily deployment of workers to bases across Afghanistan. PPI was not the only contractor that went to great lengths to get workers past Philippine authorities. Christian, who was hired as a baker by Serka in 2008, was provided a tourist visa to Turkey and booked on a byzantine string of flights that avoided Manila's international airport and departed first to Singapore to lessen suspicion of immigration authorities: "Manila to Cebu [a large city in the Visayas region of the Philippines]; Cebu to Singapore; Singapore to Dubai; Dubai to Turkey, Istanbul; Istanbul to Adana." As a further precaution the company gave Christian and twenty-four fellow passengers instructions concerning the line they should enter at Cebu's airport to assure their safe passage out of the country. Upon arrival in Adana they were put on a bus and transported overland to bases in northern Iraq.

In addition to continuing its underground recruiting efforts in the Philippines, PPI also targeted Filipinos and Indians already working in Dubai. One woman, Grace, was working under the table while on a tourist visa at the City Centre Deira shopping mall in Dubai in 2005 when she was approached by a Filipino representative from PPI who asked if she wanted to work in Iraq.

> GRACE: I said, "Iraq? They're having a problem here. They say that it's the worst going there." "No, it's not. I've been here," he told me. So I said, "OK, how much will I be earning?" He said, "For a start, they're going to give you $550 [a month]." Actually, at that time, $1 is equivalent to 56 pesos, that's why I grabbed it. I said, "OK." "After six months, they're going to increase you $50, you will be earning $600." I said, "OK, I'll get it."
>
> ME: How much more was that than you were making in Dubai?
>
> GRACE: It was triple. . . . I'm a single mom. I'm thinking, I got three daughters going into college. How can I [be] able to send them money if I don't—if I can't find a good job in Dubai? That's why instead of thinking, "Oh, Dubai is a nice place," it was, "Why not try Iraq?"

Grace recalls that PPI was looking for Filipinos with older passports that did not have a travel ban stamp (Figure 7.2), as this would make it easier to get them past Dubai airport officials onto a flight to Baghdad.

Whether through an agency or one's own initiative, a noticeable difference between legal and underground recruiting concerns how job opportunities are made known to potential workers. Advertising in print or online is out as it attracts government scrutiny. Chance encounters with crowds of applicants in front of recruiting agencies in Manila—which was how Carlos found out about PPI's hiring binge in early 2004—also ceased. Therefore following the bans, military contractors and recruiting agencies have become more dependent on local agents to inform people about hiring initiatives. Michelle, for example, facilitated the recruitment of most of her neighbors for Serka after the travel ban to Iraq was imposed. Also, as noted above, companies that have large Filipino contingents in Iraq or Afghanistan mobilize workers to spread the word about hiring opportunities to families and friends when they call home, or reach out to former employees about returning for another contract. As a result, the travel bans have served to further concentrate the pool of potential workers by ratcheting up a spatial path dependency when it comes to recruiting military labor. As one recruiting agency told me, "Once the ban is in place you can't cast your net that wide, you have to use the existing connections, which keeps it [recruiting] within the same communities that started out."

FIGURE 7.2. Philippine passport with travel ban stamp

A second distinction involves risk. While some Filipinos recruited to work in Iraq prior to the travel bans endured substandard living conditions on military bases, or were forced to work far more hours than stipulated in their contracts with no extra pay—as was the case with Serka's employees—the state provided some modicum of oversight over the recruiting process, which lessened risk. This is especially apparent when one compares the experience of Filipinos during the pre-ban period with South Asian workers recruited by dodgy and unregulated agencies like Subhash Vijay and Moon Light.

Navigating the underground recruiting process is a much more precarious proposition. To begin, successfully obtaining a job is significantly more expensive, from exorbitant fees charged by recruiting agencies to bribes paid to airport officials to smooth passage to Dubai. Moreover, there is no guarantee that the fees and bribes will result in employment. Job scams have become common. In 2012 more than twenty Filipinos were stranded in Afghanistan after promised construction jobs failed to materialize. They each lost the $1,400 dollars they had paid to a company called RMR Construction.[26] For those detained at the airport, any money paid to recruiters or contacts in Iraq and Afghanistan is lost, and they receive greater scrutiny from officials in the future. Thus from the perspective of Filipino workers, finding military work under the travel ban regime in the Philippines more resembles the experiences of South Asian laborers when it comes to risk. Consequently, according to Edward, those willing to take their chances are often more desperate: "During the ban you can only recruit who wants to go undocumented. Usually these people are really desperate to have a job. . . . You cannot hire workers with skill really. A [skilled] worker will not go undocumented, *di ba* [right]? You cannot get people who are really experienced. You can only get fast the tricycle [taxi] driver." Yet despite increased risks, the Philippines has continued to be a major source of labor for military contractors following the imposition of the travel bans. In 2010 recruiters in Manila estimated that more than 5,000 Filipinos were working in Afghanistan, with this number "growing" every day as people continued to sneak into the country.[27]

Those who successfully obtain a job can still to be negatively affected by the travel bans. Several people I talked with decided not to go home between contracts, fearful that they would not be able to leave the Philippines again. Fidel stayed for three years in Iraq even though he was suffering from lung problems that required major surgery when he came home in 2007. Those who do return often face the prospect of losing their jobs, or coming up with money to bribe officials at the airport when they leave. Rick, a firefighter in Afghanistan who was at home on vacation when I met him in 2015, described his experience for me:

> RICK: After one year, 2008, that's the time they banned Afghanistan for all the Filipinos. That's why when we go on vacation after one year of contract—we had to go on vacation—so others cannot come back because Afghanistan is banned already. We took the chance of paying money through escort in the airport so that we can exit.
>
> ME: What do you mean an escort, because other people talk about this?
>
> RICK: Some travel agency, they offer us, "If you want, I have an escort and helping you to go exit."
>
> ME: To get past immigration?
>
> RICK: For immigration. But that's on your own risk because sometimes even though you pay already and you pass the immigration, in the boarding still they checking, "Where you going?" "Afghanistan." They will pick you up and put you out again, so you spend a lot of money [for nothing].
>
> ME: How much did you spend for an escort through?
>
> RICK: The first year I go on vacation and then come back, I pay 14,000 [pesos—roughly $300], and then second it became 18,000 [$350], then the third is 30,000 [$600]. Others they pay 40,000 [$800] or a $1,000 dollars. That money when we reach Afghanistan and we got the receipt that we pay escort, the company will pay us.

ME: Wait. The company reimbursed you for the escort?

RICK: Yes. First, it come from your pocket because company they don't know if you passed the immigration. When you got there because they need you, they need workers, that's the time they will reimburse the money you spent in escort and everything . . . as long as you have receipt they will pay.

Rick was the only person I talked with whose employer reimbursed him for the expense of evading the travel bans. His situation is unique due to the fact that as an airport firefighter he had specialized skills that were in high demand, and because his company is a prime contractor for the military. As I discuss more in chapters 8 and 9, this means that his status and privileges are much greater than those working for subcontractors.

While the travel bans pushed recruiting and travel to and from bases underground, thus increasing workers' precarity, Filipinos I interviewed also demonstrated remarkable initiative and agency in obtaining work with military contractors during this period. Take Anne, who was fired by Serka in 2006 for participating in protests against restrictive rules forbidding cell phones and the ability to move around her base unescorted. Two years earlier she had joined a large labor strike after Serka promised workers $600/month (which was stipulated in the contracts she and others signed) but paid only $300/month for DFAC work. At that time she and other strikers were supported by KBR and the military. But her participation in protests two years later—against rule changes that were introduced by the military, not her employer—convinced Serka to terminate her contract and send her home. Not content with staying in the Philippines, Anne procured a tourist visa to Dubai and "approached every agency in Dubai" about work in Iraq. After seven months searching, an Indian-run agency offered her a job with Kulak. Another laborer, Andrew, returned home in 2008 after four years with PPI in Iraq. Then in 2009 he decided to look for military work again. He contacted a Filipina fixer living in the UAE who was recommended by a former colleague in Iraq. The fixer helped him obtain a tourist visa and plane ticket to Dubai, put him up in a crowed apartment—"Just a two bedroom, [with] 40 Filipinos"—and arranged interviews with military contractors. Three months and $2,000 in fees later he obtained a job with a U.S. engineering firm, Arkel, in Afghanistan.

Another option involves smuggling oneself into Iraq or Afghanistan, with the hope of finding a job with a military contractor after you arrive. Two examples suffice. The first comes from Mary, who smuggled herself into Victory Base Complex in Baghdad in 2005 after a promised job that she paid $1,400 for fell through.

MARY: We got to Dubai [and] they told us that they don't hire to Iraq . . .
I'm crying, always crying, because I am thinking of my mother's land.
It's in the pawnshop [mortgaged]. Then someone told me, "You want
to go to Iraq, one week, [by] cargo airplane?"

ME: With a contract or without?

MARY: We don't have a contract. Three girls. We go there. We don't have
a visa, we don't have a contract. . . . We go to the [Baghdad] airport.
We go into the back where somebody pick us up by bus.

ME: How did you find out about the bus?

MARY: My coworkers [friends] in PPI. They were already there.

ME: PPI didn't give you a contract but your friends said, "If you come
with us we can find you a contract, we'll get you a job"?

MARY: Like that, yeah . . . One month before we work, we go outside.
[Until then] we only stay in the [PPI] camp. If you don't have badge,
you don't go outside the camp. It's only in the lady's camp we stay
there.

Like Mary, John's cousin initially smuggled herself into Afghanistan without a
job offer. John explained how this works for me.

JOHN: There is a camp right outside of KAF [Kandahar Airfield] that is
an Afghan [army] camp, and Filipinos often stay there to look for
jobs.

ME: So you would fly into KAF, leave the base, and then look for a
job?

JOHN: You had to coordinate with relatives or friends to find a job,
because you wouldn't be able to get back on the base without papers
[John is referring here to a letter of authorization (LOA), which states
that the person is an employee of a contracting firm and thus authorized to be on the base].

ME: How long would people stay there?

JOHN: A month. Five months. Sometimes people would stay that long
because they couldn't find work. Sometimes they went by land from
Dubai, because the food comes from Dubai through Pakistan [and]
they hitch a ride.

Mary's and John's cousin's stories are but two of many examples of Filipinos sneak-
ing into Iraq and Afghanistan for work that were recounted for me. Given the
widespread flouting of travel bans by the military and its contractors, it is understandable that individuals are willing to risk smuggling themselves into an active
warzone in the hope of finding work.

Nearly every news account of the experiences of South and Southeast Asian laborers working for the U.S. military in Iraq and Afghanistan focuses on trafficking and labor abuses such as wage theft. That the military has done little to combat these human rights violations by its contractors and subcontractors is disgraceful. But there is a danger in this exclusive focus on exploitation, in that TCNs tend to be painted as passive and helpless victims. The above stories offer a necessary corrective to this view. Indeed, time and time again I was struck by the ingenuity and courage displayed by Filipinos who have labored in Iraq and Afghanistan. This is especially the case when it comes to labor activism by workers like Anne, which is the topic of the next chapter.

Evading Responsibility

In 2008 families of the Nepalese workers killed in Iraq filed a federal lawsuit in Texas alleging that the men were victims of a trafficking scheme organized by Daoud & Partners and KBR. After initially allowing the case to go forward, the judge reversed course in 2014, declaring that congressional anti-trafficking legislation "was silent with regards to extraterritoriality" prior to 2008, therefore the plaintiffs did not have standing as the alleged crimes did not take place on U.S. soil.[28] In similar fashion, a 2006 DoD investigation into the incident declared that "the U.S. government had no jurisdiction over the persons, offenses, or circumstances that resulted in the Nepalese deaths."[29] On its face this was a rather curious claim given that Daoud & Partners was one of the largest military subcontractors in Iraq and the men were kidnapped while in transit to work on U.S. bases in the country. Such comments, however, are consistent with the desire of the military to distance itself as much as possible from responsibility for an oversight role when it comes to subcontractors' labor practices.

Indeed, what stands out in the Najlaa and Nepalese cases—as well as many other incidents—is the inability or unwillingness of the military to provide effective oversight of subcontractors with regard to their treatment of workers, despite the fact that these workers' employment is entirely a consequence of military contracting. While acknowledging legal complexities raised by overseas contingency contracting, it is striking how often narrow, legalistic arguments are used to justify or explain lack of oversight responsibility when it comes to labor exploitation by subcontractors. Following its initial investigation into the Najlaa affair, for instance, the Defense Contract Management Agency (DCMA) concluded that "the USG [U.S. government] does not have jurisdiction over these TCNs, as these men are not being held on USG property, nor do they have USG contracts."[30] More farcically, a DoD Inspector General report argued that "while certainly disconcerting, the facts and circumstances did not suggest that Human Trafficking Violations had occurred" because "TCN personnel housed in the . . . complex were free to leave if they had decided to do so."[31] Move along, nothing to see here, in other words.

So how do military officials justify evasion of oversight responsibility? One way, as noted above, is by invoking the legal principle of privity of contract. According to privity doctrine contracts establish a legal relationship between parties, with attendant rights and obligations. This relationship does not apply to third parties. Thus in the case of defense contracts, the government and prime contractors have privity, prime contractors and their first tier of subcontractors have privity, but the government, as a third party, does not have privity with subcontractors. In other contexts the U.S. government has typically invoked privity of contract to shield itself from claims by subcontractors.[32] In fact, it is against government policy to deal directly with subcontractors in order to maintain the legal distance that privity provides.[33] But privity can be permeable going the other way as the government has a variety of tools available to monitor and enforce subcontractors' policies and behavior if it so desires.

The most powerful tool is the use of "flow down" clauses. This involves directing prime contractors to insert clauses into contracts with subcontractors requiring the latter to comply with certain provisions. For example, a standard "audit clause" requires a subcontractor to allow the government to examine records of cost and pricing data.[34] Federal regulations also now require a "combating trafficking in persons" clause for all contracts, including flow downs of this clause for subcontractors. The current version of this clause prohibits the use of forced labor in the performance of contracts, confiscation of passports, use of recruiters that do not comply with labor laws of the country in which recruiting takes place, and the charging of recruiting fees. It also directs companies to "provide timely and complete responses to Government auditors' and investigators' requests for documents" and "reasonable access to facilities and staff . . . to ascertain compliance" with prohibitions against trafficking.[35] So in theory, at least, privity should not be a significant stumbling block to effective oversight.

In practice, however, the military's efforts to enforce prohibitions against trafficking through contracts amount to little more than legalistic formality. Black and Kamat describe the experience of one former worker in Iraq who recalled that upon arriving at the military base "his contractor required him and his colleagues to sign a Trafficking Awareness form, issued by the Department of Defense. 'We all knew—and they knew—that we had paid,' he said, referring to his supervisors. 'Oh, yeah, everybody knows.'"[36] As this example illustrates, the military does not assertively ascertain whether or not trafficking has occurred, but effectively outsources this task to contractors, who are tasked with developing and implementing a "compliance plan."[37] The problem here is that "contractors essentially have been asked to turn themselves in upon learning that an employee has violated this policy—even at the risk of contract termination, suspension and debarment. Thus, while the FAR [Federal Acquisition Regulations] and DFARS [Defense Federal Acquisition Regulation Supplement] ban on human trafficking is a warning to Contractors that such activities are expressly prohibited, it is doubtful that the regulations will accomplish their laudable objectives, since Contractors are unlikely to self-report."[38] Moreover, prime contractors, who are tasked with policing the behavior of their subcontractors, are largely dependent on the latter's cooperation, especially when it comes to conducting interview checks with workers. One Fluor employee from Bosnia who worked as a QA/QC supervisor in Afghanistan explained to me: "Usually they [subcontractors' foreign employees] don't speak English at all. I had a language assistant which is from their company. That was against the contract . . . in my documentation I'm not supposed to have any person next to me and especially from [the] same company. There's no other way, though. I had to have one. Your [the subcontractor's] language assistant between me and him." From the perspective of workers this pro forma process provides no incentive to speak truthfully about their experiences. As an Indian worker for a subcontractor in Afghanistan told Black and Kamat, "We've already paid agents for the job. If we tell the U.S. military that we paid a fee, they'll just send us back, and we'll lose everything."[39] Given all this, it is not surprising that a 2011 report on wartime contracting commissioned by Congress concluded, "The Commission uncovered tragic evidence of the recurrent problem of trafficking in persons by labor brokers or subcontractors of contingency contractors. Existing prohibitions on such trafficking have failed to suppress it."[40]

The second, and primary, legal justification against more robust enforcement of anti-trafficking prohibitions and other labor standards centers on jurisdiction, or rather a supposed lack thereof. Westphalian sovereignty is based on the assumption that political borders define jurisdiction, with each state possessing absolute authority to enforce the law within its territory. In reality this territorial ideal has always been riddled with extraterritorial exceptions where domestic law extends beyond borders. This was particularly true during the age of empire in the late nineteenth and early twentieth centuries. John Darwin goes as far as to claim that "extraterritorial 'rights'" ensured by "bases, enclaves, garrisons, gunboats, treaty ports and unequal treaties" were "as much the expression of . . . European imperialism as were the colonies and protectorates."[41] Nor was extraterritoriality limited to European powers. By 1900 the U.S. signed a number of treaties guaranteeing extraterritorial jurisdiction over its citizens living in North Africa (Morocco, Algiers, Tunis, Tripoli), the Middle East (Turkey, Muscat, Persia), Asia (Japan, China), and several other locales. The most extensive of these was China, where Americans, along with European foreign residents, "enjoyed virtual immunity from native law, and were instead under the extraterritorial authority of their own home governments."[42] By the turn of the century legal demands created by the large American presence in China prompted the establishment of a "U.S. Court for China" based in Shanghai, which operated until 1942. "In sum," legal scholar Kal Raustialia observes, "empires and extraterritoriality were closely linked."[43]

Echoes of these imperial extraterritorial exceptions continue, most significantly SOFAs that provide for varying degrees of extraterritorial jurisdiction over U.S. personnel and dependents deployed to overseas bases. In 2000 Congress passed the Military Extraterritorial Jurisdiction Act (MEJA), which closed a jurisdictional "legal Bermuda triangle" by extending extraterritorial authority for certain crimes over military contractors.[44] Notably, MEJA applies to foreign nationals as well as U.S. citizens. Unfortunately, MEJA's effect has been minimal. In the first ten years after the law's passage only fifteen attempted and successful prosecutions involved civilian contractors—and none of these concerned trafficking or other labor abuses.[45] One reason for this is the difficulty U.S. prosecutors face in gathering evidence overseas, especially in war zones. More significant, though, is "a simple lack of political will to bring cases."[46]

This lack of will is often masked by specious references to jurisdictional gaps that no longer exist. The Najlaa case is emblematic here. Nowhere in the MEJA, for instance, does it state that extraterritorial jurisdiction is limited to crimes that occur on U.S. property or bases, as DCMA claimed. In fact, MEJA was successfully used to prosecute the four Blackwater contractors who killed fourteen civilians at Nisour Square in Baghdad in 2007. Najlaa, in contrast, was never prosecuted for its labor abuses. It even continued to receive contracts from KBR and the military. This despite the conclusion of officials at the U.S. embassy in Baghdad that the incident was "essentially the trafficking of low-skilled expat workers into forced labor" due to the fact that "these people are only making $300 to $400 a month (for 12hrs/day 7day work weeks) and they are effectively working for little or nothing for the 6–12 months it takes them to recoup the broker's fee."[47]

In addition to MEJA prosecutions the military has a number of other legal and policy avenues at its disposal if it wished to be more aggressive in curtailing labor abuses by subcontractors. One possible step, recommended by the Commission on Wartime Contracting in Iraq and Afghanistan (CWC), would be to "require that foreign prime contractors and subcontractors consent to U.S. jurisdiction as a condition of award of a contract or subcontract," thereby eliminating any potential confusion as to the reach of U.S. law.[48] Another approach would be to use the Uniform Code of Military Justice, which gives the military jurisdiction over civilian personnel working with troops in overseas operations, to prosecute contractors for trafficking or other labor abuses.[49] Beyond prosecutions, the military could also be more aggressive in pursuing debarment or suspension of contractors in response to evidence of trafficking and other labor abuses. That it does not pursue these avenues or MEJA prosecutions is telling. Especially when one considers the lengths to which the U.S. government and courts have expanded extraterritorial jurisdiction across a range of domains in recent decades—from the "war on drugs" to foreign sovereign debt disputes.[50] Indeed, it is hard to disagree with the conclusion of one legal analysis of trafficking by military contractors that "the main issue plaguing the U.S. Government in preventing and prohibiting human trafficking is, predominately, the Government itself."[51]

This problem, I believe, ultimately stems from the fact that officials are unwilling to acknowledge that contracting out logistics support to an offshore army of workers means that the military is responsible for the conditions under which this workforce is acquired and labors on its behalf. In this the military is similar to large U.S. corporations with extended offshore supply chains that try to evade responsibility for substandard labor conditions suffered by workers at the end of these chains. In both cases offshoring labor has transformed the workforce, from American to foreign. The consequence, as Maya Eichler observes, is that military contracting—like offshore manufacturing—depends on and reinforces global inequalities of citizenship that intersect with racial, class, and gendered inequality.[52] When this is combined with subcontracting and extended labor supply chains that downsource risk and attenuate moral responsibility, it is not surprising that U.S. civilian and military officials' response to cases of trafficking and other labor abuses has been so tepid.

Part 3
BASE LIFE

ACTIVISM

We were all thinking the same, so in the evening we talked and agreed that we would not go to work in the morning. . . . We called a meeting. One or two men went room by room. The guy who led it was a former OFW, so he had lots of experience. So he went room by room and convinced people. When he came in he said, "Tomorrow we will do something about the salary. Please join us. It won't take long."

—Manny

Third country nationals have a precarious and liminal status on U.S. military bases in the Middle East and Afghanistan. In contrast to American contractors, they have little recourse to remedies through the U.S. legal system when subject to labor abuses, especially when their employers are also foreign firms, as most military subcontractors are. And unlike local laborers who can appeal to host countries for help, they have few external political and social relationships that can be mobilized to advocate for better wages or working conditions. Moreover, military bases are essentially closed company towns where all workers are at-will employees that can be fired and deported by the military or contracting companies for any reason. In many cases being deported results in being "blacklisted" by the military from working again on its bases.

Given this context, U.S. bases in warzones are one of the last places one would expect to find a ferment of labor activism. This impression is buttressed by the fact that few news stories discuss labor strikes or protests on bases. In part this is a product of restrictions governing where reporters can go and who they can talk with. But it also reflects a widespread narrative that portrays TCNs—especially those from South and Southeast Asia—as hapless victims with little agency of their own. One notable exception to the lack of reporting on labor activism is the work of Sarah Stillman, whose excellent 2011 *New Yorker* article, "The Invisible Army," describes a massive 2010 riot by PPI workers at Victory Base Complex in Baghdad. Angered by a shortage of food at the company DFAC, more than 1,000 workers, mainly from India and Nepal, ransacked their mancamp, "smashing windows, hurling stones, destroying computers, raiding company files, and battering the entrance" after supervisors refused to provide more rice. Eventually U.S. military police and Ugandan security guards were called into the camp to quell the riot. Several weeks later GCC employees in Baghdad staged their own protest, "pelting their bosses with stones and accusing the company of failing to pay them their proper wages."[1]

As Stillman's article hints at, protests and strikes have in fact not been uncommon on bases in warzones. Nor are these the only forms of labor activism. More common, I am told, is a practice that workers refer to as "jumping," which involves surreptitiously transferring from one company to another in search of better pay or working conditions. As I argue below, this is a strategy that is no less risky than mass protests or strikes.

Viewed in a broader context, these labor struggles can be situated within a long history of such activities at the myriad of overseas U.S. military projects and bases. Julie Greene describes repeated strikes, riots, and attempts to organize unions on the part of the multinational workforce the military enrolled to construct the Panama Canal in the early 1900s—as well as various coercive measures introduced to repress these activities.[2] Labor unrest also plagued contractors in Vietnam. In 1966 roughly 16,000 RMK-BRJ workers at multiple bases struck for better wages.[3] That same year 4,300 Korean, Filipino, and Vietnamese laborers for RMK-BRJ at Cam Ranh Bay military base went on strike in protest of onerous work rules.[4] And in late 1967 some 2,000 Koreans working for Vinnell Corporation at Cam Ranh Bay protested a shortage of rice and the quality of food they were being fed. When a company manager shot three Koreans, the protests turned into a multiday riot, with workers smashing bulldozers and trucks into buildings.[5] U.S. bases in the Philippines were a regular target for labor activists prior to their closure in the 1990s. In 1986 more than 22,000 Filipino workers struck at Clark Air Base, Subic Bay naval facility, and six smaller bases. Seeking pay raises and increased severance benefits they blockaded entrances with logs, rocks, and scrap metal for nearly two weeks, preventing service members from entering or leaving the bases.[6] More recently, in 2013 hundreds of Djiboutian laborers conducted more than a month of protests and strikes against planned workforce cuts by KBR at Camp Lemonnier, forcing troops to "man the chow line" and perform other logistics duties.[7]

There are several factors that make labor activism by foreign workers in warzones distinct from cases like the Panama Canal, Djibouti, and the Philippines. One is that in the latter local labor laws govern—to greater and lesser degrees depending on SOFAs and other bilateral agreements—hiring and firing procedures, wages, rights to unionize or strike, and working conditions. Foreign workers have no such legal frameworks that they can effectively appeal to. Another is that as transnational labor migrants TCNs are a captive labor pool whose alternative in-county work options are essentially nonexistent. Being forced to return home is the most likely outcome if they leave, or are terminated from, employment on a base.

Perhaps the most significant difference involves the political and social dynamics of protests and strikes. In countries that host large, long-term U.S. bases the military is highly reliant on local labor to provide logistics support. This dependence, Amy Austin Holmes argues in her analysis of domestic social unrest connected to military bases located in Germany and Turkey, means that local workers possess a degree of "structural power." By this she means the ability to use strikes and other forms of labor unrest to further various economic and political goals, ranging from obtaining wage increases and greater job security for workers to pressuring the host government to end the U.S. military's presence. Holmes's research demonstrates that the ability to mobilize structural power is not just a matter of labor dependence; it is also shaped by a triadic relationship between the military, local labor, and the host nation. Thus in responding to strikes and protests the military by necessity has to take into account how its actions will be received by the host society and its political elites.[8] Such considerations are absent in the case of TCN labor activism in the Middle East and Afghanistan, greatly attenuating the structural power that these workers possess. Put another way, the use of foreign workers in warzones produces a relatively high degree of "internal containment" of labor relations. This, as noted in chapter 7, is a desired feature of labor offshoring, which facilitates the military's goal of developing more flexible and modular logistical support regimes.

I do not wish to imply that labor struggles on these bases unfold as if trapped in a hermetically sealed container. A certain amount of political leakage is inevitable. In May 2005, for instance, some 300 Filipino employees of PPI went on strike at Taji Air Base (also known as Camp Cooke at the time). They were later joined by 500 laborers from India, Nepal, and Sri Lanka. The strikers accused PPI of violating contract language concerning working conditions and hours, falling three months late on pay, and refusing to provide cooks in the company man-camp that would prepare national dishes such as adobo chicken. In response PPI threatened to immediately fire the agitators and send them back home on chartered flights. When word of the labor dispute was picked up by Philippine diplomats in Iraq, the government dispatched its chargé d'affaires, Ricardo Endaya, to mediate talks between PPI and the workers. These talks ended in a quick settlement. Endaya also investigated allegations of systemic labor abuses by First Kuwaiti in 2004 and 2005—there were "simultaneous complaints at multiple camps" he recollects—but the company refused to meet with him to discuss remedies. During his time in Iraq, Endaya continually prodded U.S. officials and his superiors to more aggressively investigate First Kuwaiti's labor practices, though to little effect.[9]

This chapter explores the hidden phenomenon of labor activism on U.S. bases in Iraq and Afghanistan. Foreign workers on these bases have three choices when it come to their situation. The first, and safest, is a "don't rock the boat" approach. That is, keep one's head down and continue to work without complaining or trying to change conditions. A second option is to return home. The problem is that this is a road to economic ruin if one paid exorbitant recruiting fees and still owes money to loan sharks, as many workers do. Finally, workers can decide to engage in labor activism—either collective or individual—aware of the risks that this entails. In the rest of the chapter I examine when, why, and to what effect workers choose this latter option. I also discuss contracting companies' strategies to suppress workers' efforts, and the military's ambiguous position in relation to these struggles.

Protests and Strikes

In 2004 strikes unfolded across several major bases in northern Iraq, beginning at Diamondback and Marez in Mosul, and then spreading to Q-West and Tal Afar. Instigated by Filipinos working for Serka, which held DFAC contracts for the bases, these actions were one of the earliest examples of large-scale labor activism in Iraq. They were also successful, leading to significant pay increases for strikers at all four bases. I first heard of the strikes from Daniel, a gregarious former Serka employee who was one of the first people I met in the Philippines. At the end of the interview I was left with several questions, the answers to which only started to come into focus as I talked with other participants in these strikes. Why were the majority of Serka's Filipino workers motivated to take action despite the obvious risks? How did the process unfold? And what were the circumstances that contributed to a favorable outcome for participants?

More than a decade after the strikes occurred Daniel still relished describing the events and displayed evident pride in what he and his fellow workers accomplished. Serka's Filipino employees had two main complaints with their status. First, they were angered by the gross discrepancy between contracts signed in the Philippines and pay and working conditions in Iraq. As mentioned in chapter 7, Serka's contracts specified an eight-hour a day work schedule, with one day off each week. When people arrived in Iraq they were told that standard shifts were actually twelve hours, with no days off. And despite the extra thirty-six hours a week they were working, they would receive no additional pay. In addition to this blatant wage theft, many workers found themselves paid far less than promised.

Anne, for instance, was promised a salary of $600/month, but was paid only $300. Other Serka employees also mentioned working for salaries far less than promised. Their accounts are bolstered by a 2006 Army investigation into Serka that found evidence that this practice continued after the travel ban drove recruiting underground.[10]

What Filipinos really resented, however, was the fact that Serka's Turkish employees were being paid much more for doing the same work. Manny recalls, "A Turkish worker only go clean the toilet but his salary is $1,000 [a month]. A Filipino that got the same [salary] work on the computer. For example, when I was a driver my salary was $1,000. My helper, Turkish, was $1,300 to $1,400 . . . Turkish workers in the kitchen would get twice as much [as Filipino kitchen workers]. . . . We felt insulted by the Turkish salaries compared to ours." Moreover, Filipinos felt that the company's Turkish employees treated them shabbily, even though they were not as qualified. Daniel remembers, "We felt disrespected by the Turkish workers. [They] didn't go to school. They didn't know how to write, how to read. And they didn't know English . . . But they were very full of themselves." Serka is a Turkish company with a history of providing logistics support for U.S. military bases in Turkey going back to the 1960s. Therefore the fact that it treated its compatriot workforce better than its Filipino employees makes sense, to a degree. In this, in fact, its practices resembled Fluor's and DynCorp's tiered distinctions between American, European, and Asian workers in Afghanistan a few years later. But Filipinos I talked with didn't see it this way. While the racialized hierarchy between American workers and TCNs is generally internalized and accepted by Filipinos—especially if the latter work for a subcontracting company—they objected to the assumption that Serka's Turkish employees should have a higher status.

Daniel suggests this perspective was fueled by discussions with Filipino-American soldiers, who said to him and others, "You guys are getting screwed." In his telling these conversations planted the seed for organizing against the company. Rodrigo, one of his coworkers at Marez, says that the first step was drafting a petition that was delivered to Serka and KBR management. "They drafted a letter that said, 'We're asking for [an] increase. We know how dangerous the job is. You know how dangerous the situation is which makes our jobs dangerous as well. Here are the list of the names who are demanding this and they signed them. If you cannot give this to us, send us home.'" According to Rodrigo, approximately half of Serka's Filipino workforce on the base signed the petition. The following day they stayed in their barracks instead of reporting for work at the DFAC. Almost immediately the base mayor called a meeting with Serka, KBR, and strike leaders. According to Daniel, Rodrigo, and Anne, both KBR and the military backed their demands, with KBR telling Serka's managers, "Don't make this get any worse, fix it." Once people heard this, they knew they had won. Three weeks later Serka doubled the salary for all of its Filipino workers on the base, even those who did not participate in the strike.

In short order strikes spread to other bases in northern Iraq. The last base to be organized was Tal Afar. Manny, who worked there, was originally reluctant to join the strike: "I was scared. On the one hand I wanted my salary to be increased. On the other I was scared I would be sent home. The one consolation I had was that if I got fired everyone would get fired." When I asked if others felt this way he said: "Yeah, many. They were scared too. But we were [also] afraid we would be teased if we didn't [go along]. . . . Also, we heard about strikes in other camps. We heard that other camps were able to negotiate their salary, so we thought, 'Why don't we try?' Hearing about it [the other strikes] the leaders were confident that it [a strike] would work." In the end Manny and most of Serka's non-Turkish workforce at Tal Afar, which included some Indian and Egyptian workers, joined the strike.

Several factors contributed to the success of these strikes. As mentioned above, the perception of injustice concerning Serka's differential treatment of Turkish and non-Turkish employees created a shared set of grievances ("we were all think-ing the same") that helped unite its Filipino workers against the company. In addition to this, it is clear that those with previous experience working abroad, such as Daniel and the leaders of the strike at Tal Afar, played an important role in mobilizing people. Also relevant is the power of demonstration effects. Ac-cording to Manny, success at the bases in Mosul emboldened both leaders and reluctant followers like himself to undertake their own strikes. Perhaps most important was the fact that KBR and military officials sided with Serka's Filipino workers, which foreclosed the possibility of punitive actions against participants by the company.

This support has to be understood in relation to the specific dynamics at play. It is not immaterial, for instance, that Filipinos constituted the vast majority of Serka's food service workers at the time. Without their labor the company lacked the manpower to fulfill its contractual obligations. Moreover, the military places great emphasis on dining operations in warzones, believing that plentiful and high-quality food is critical for maintaining morale. Food service is also much more time sensitive than other logistical support activities like construction, main-tenance, and transportation. Truck convoys can be delayed for days with little disruption to operations, but if troops miss a meal because the DFAC is shut down all hell breaks loose. And due to strict sanitation rules and procedures, it is dif-ficult for companies to rapidly replace striking food service staff with untrained workers. Combined, the constant rhythm of food service and the critical mass of Filipino DFAC workers who went along with the strike meant that strikers had a significant degree of leverage. Therefore from the perspective of KBR and the military the quickest way to resolve the problem of interrupted food service was to order Serka to "fix" the issue by meeting its workers' demands.

In response to the strikes, Serka made several labor management changes. First, it appears to have phased out hiring Turkish workers, whose elevated status generated so much resentment from Filipinos. Second, the company became more aggressive in punishing employees who tried to organize protests or strikes. As we saw in chapter 7, Anne was deported in 2006 for protesting new restrictions on the use of cell phones and freedom of movement around her base without escorts. A critical difference here was that these restrictions were imposed by the military, so when Anne protested the changes she received no outside support. But rather than explain the situation for Anne and other protesters, Serka took this as an opportunity to remove unwanted agitators from its workforce. Another Serka employee, Angelo, recalls protests and strikes over low wages by Indian and Bangladeshi workers in 2006. Initially the company agreed to increase their $300/month salaries by $50. But when dozens of workers continued to complain about this small increase in pay, Serka rounded up the leaders and sent them home.

The key change introduced by Serka was diversification of its workforce by nationality. According to Daniel, "In the beginning it was mostly Filipinos and Turkish, but after the strikes Indians and Nepali came in." These South Asian workers, he claims, cost less money and were more docile. This shift in workforce composition, which I call the "Tower of Babel strategy," was deliberate according to those I talked with. The logic behind diversifying one's workforce is that it is harder to organize and mobilize support for mass action across national lines. Articulating shared goals or strategies, for example, is more difficult with a multinational workforce, especially given the fact that most South Asian workers have little to no knowledge of English. Moreover, levels of solidarity and trust across national lines are much lower than within. For example, when I asked Angelo why he and other Filipinos refused to join the strike by Serka's Indian and Bangladeshi employees he replied, "If you complained about salary you could get fired" and "it was their own strike." Serka played up these tensions by paying different wages based on nationality. Filipinos on Angelo's base were paid double the company's South Asian workers, for instance. More importantly, a heterogeneous workforce allowed Serka—and other companies that pursued this strategy—to more effectively marginalize labor activism, especially if it remained largely confined along national lines. When a DFAC workforce is constituted by a mix of Nepalese, Indians, Bangladeshis, Filipinos, and Turks, protests or work stoppages by one or two of these groups can be more easily absorbed without dramatically affecting operations. According to Angelo, this was why the 2006 strikes met with limited success. Even though a significant portion of both Indian and Bangladeshi workers participated, they still represented a fraction of DFAC workers, which decreased their bargaining leverage. Another Serka employee, Christian, recalls a 2009 strike by fifteen Bangladeshi DFAC employees. Rather than negotiate with this small contingent of disgruntled workers the company immediately sent them home, with little visible effect on food service.

Filipinos who worked for other subcontractors, including the military's largest body shops in Iraq, PPI, GCC, and Kulak, suggest that this approach to workforce composition, and aggressive moves to identify and deport labor activists, quickly became widespread strategies for suppressing mass protests and strikes. Despite this, interviews reveal that labor disruptions continued to occur at bases across the country. One PPI employee who was at Camp Bucca in 2006 and 2007 recalls a large work stoppage where "the majority" of the company's employees, "people from India, Philippines, Nepal," refused to work for two days in protest of small salaries and lack of overtime pay. Built in the desert wastes northwest of Iraq's main port, Umm Qasr, Bucca was the U.S. military's largest detention facility in the country, holding 26,000 Iraqi prisoners at the peak of its operations.[11] After promising a small salary bump, PPI managers convinced everyone to return to work. A week later the company terminated eight employees—six of them Filipino—that it identified as the strike leaders.

Another person I talked with, Chris, remembers a labor protest in early May 2005 staged by approximately 500 Filipino GCC employees at Victory Base Complex in Baghdad. Unlike the cases discussed above this one concerned food. Specifically, the fact that the company dining hall was staffed with Indian cooks who refused to accommodate their requests for noncurry dishes. "They fed us with curry that looked like shit. There were so many people [that] got stressed because of that." After complaints to company managers went nowhere, Chris and the rest of GCC's Filipino contingent turned to the military for help. "We went to the staging area, the military staging area and said, 'We refuse to work.' One officer came up and said, 'What is the problem with your company? I will call your company [managers].' 'Sir, our problem is food, not work. Food, only food.' The military gave us some MREs and water and we went back to work. . . . What KBR did, [is] they talked to the [GCC] management to fix the problem. They provided us food accommodation, Filipino food."

This is an instructive example of labor activism for a couple of reasons. One is that it illustrates another issue—the type, quantity, or quality of company-provided food—that frequently motivated workers to protest or strike. Several people discussed protests that centered on food occurring with other subcontractors during the first few years of operations in Iraq, an unintended consequence, perhaps, of hiring nationally heterogeneous workforces. Eventually, at most bases the largest subcontracting firms began providing a range of food options in their mancamps to accommodate the tastes of workers from different countries. However, as Sarah Stillman's reporting discussed above indicates, food complaints remained a key precipitant of protests and strikes throughout the occupation of Iraq. Another common driver of labor activism concerns salary arrears. PPI, for example, was notorious for its frequent delays, with people I talked with going up to four months in a row without pay. PPI's salary arrears was one issue, in particular, that had the power to unite all workers, no matter their nationality, in protest.

The other significant detail about the GCC protest Chris recounted is that workers took their complaints directly to military officials after receiving no response from company managers. This too was not uncommon. Enrolling the military in labor disputes could take different forms, from large protests like the one described by Chris to discussing problems with friends or acquaintances in the military, as was the case with Daniel, who credits Filipino-American soldiers with convincing him and others at Marez to organize the first strike against Serka. Daniel was not alone in thinking that the military was more responsive than company managers and KBR supervisors. In 2004 a number of PPI workers at Balad began agitating for better food and housing accommodations, eventually formulating a list of demands. When I asked Isko, who worked there at the time, whether they presented their demands to their PPI or KBR managers he responded, "Not the mangers. They gave it to the MPs [military police]. Because the MPs had the power to change the rules or address the problem. Even if you complained to the managers there was nothing much that they would do. That's why you went to the MPs."

We should not place too much stock in these accounts of military support for labor activism. For one, most of the examples people told me about involved interventions by individual soldiers that had personal connections with workers, often developed through shared activities (church and basketball being two of the most common) or heritage in the case of Filipino-American soldiers. Moreover, it would be inaccurate to say that the military as an institution has a pro-labor disposition on its bases. Workers from Bosnia and the Philippines told me they could not recall a single case of military officials intervening to block the deportation of labor activists by either prime contractors or subcontractors. Additionally, the military's reaction to labor abuses on its bases over the past two decades has consistently been reactive rather than proactive, demonstrating—as discussed in chapter 7—much reluctance in providing substantive, systemic oversight. Nonetheless, that workers often directed their complaints to the military, viewing this as the most promising avenue for remedies, is revealing about the state of labor relations on bases.

Interviews and news accounts suggest that large-scale protests and strikes may be less prevalent in Afghanistan than Iraq. Why exactly this is the case is unclear, though several possible reasons come to mind. First, several people who have worked in both countries suggest that subcontractor salaries in Afghanistan are typically higher than Iraq. In Afghanistan, PPI, for instance, paid monthly salaries ranging from $1,200 to $1,400, approximately double what people earned in Iraq. Additionally, LOGCAP prime contractors in Afghanistan—DynCorp in particular—appear to be directly employing a greater percentage of their logistics workforce, hiring Filipinos, Kenyans, Indians, and Bosnians on "Asian" contracts. Not only do workers with these contracts usually earn more than those employed by subcontractors, their food and accommodation also tends to be of a higher standard. It is also possible that subcontractors have become more effective in surveilling and disciplining workers that engage in labor activism.

Though less prevalent, protests and strikes still take place in Afghanistan. One of the more remarkable examples was relayed to me by Rick, who works as a firefighter at KAF. One of the two main logistics hubs in the country (Bagram is the other), KAF's massive airfield plays a critical role in ferrying equipment and personnel into and within Afghanistan. In 2007 a Canadian company, ATCO Frontec, was awarded several contracts to provide support services at KAF, including fire and crash rescue. One of the countries it turned to for its labor needs was the Philippines. The company paid Rick and other Filipino firefighters $1,000 month, more than he previously earned working at Clark International Airport in the Philippines, but a pittance for such highly skilled and important work in a warzone. In 2009 ATCO Frontec's entire Filipino contingent began demanding salary raises. "We decided if they don't increase our salary we will go home," recalls Rick. "We were eighteen people at that time. We wrote a letter to the company that every year we want a raise to our salary . . . They don't answer our letter. Four letters we sent to them, nobody answered us. . . . We talked to our fire chief, 'How about the letters we sent to the company?' 'Nobody answered,' he said." After months of being ignored, they decided to escalate matters when the project manager returned from a visit to company headquarters in Calgary.

> RICK: We packed all of our things, packed our bags . . . We told the fire chief that at 5:00 we will stop working. So the fire chief called the project manager, "We got a problem here. People don't want to work. It's 5:00 so you have to come here and make a decision about these people. There are eighteen Filipinos going home if you don't answer and attend [to] their grievances." The project manager came. "This is your new contract," he told us. We read it. It was the same contract. We told him, "We will never sign this. We will leave." We went out and the project manager called us back. "What do you want?" "A

salary increase." "How much do you want?" "Double our salary." "OK," he said, because he saw that we were ready to leave and then the category of the airport will go down. This will be a big problem for the company.

ME: What do you mean by "category of the airport"?

RICK: That's civil aviation policy. It's [KAF's] supposed to be a 9, it will go to 7 [if we walk out]. The company will be fined a million dollars an hour. Because this will be reported to the tower, the Base Operation Center.

Rick is referring here to the International Civil Aviation Organization (ICAO)'s Rescue and Fire Fighting Service (RFFS) requirements. Based on a 10-point scale these categories represent assessments of an airport's firefighting and rescue capabilities. They also determine the size and type of aircraft that an airport is permitted to handle. Category 9 airports, for instance, are allowed to receive planes up to 250 feet in length, while category 7 airports are limited to planes that are less than 161 feet long.[12] The largest military transport and refueling planes, such as the C-17 Globemaster III, C-5 Galaxy, and KC-10 Extender, require category 9- or 8-rated airports. So too do the most popular civilian transport planes like Boeing's 747/767/777 variants, and Airbus's A330/340/350 series. A drop in KAF's ICAO rating from 9 to 7—which Rick claims would have been the result of a decrease in the airfield's firefighting contingent by eighteen workers—would have crippled logistical operations for southern Afghanistan. And ATCO Frontec would have borne responsibility for this ratings drop, a breach of contract that in turn would have resulted in large fines from the military.

Eighteen firefighters banding together and instigating a dramatic work stoppage showdown with their project manager on the tarmac in Kandahar: it was a remarkable event, one that won Rick and his colleagues a significant pay raise from their employer. Their actions demonstrated a sophisticated understanding of logistics operations, military contracts, airport regulations, and bargaining leverage. More significantly, in my view, this threatened strike—as with all the other strikes and protests that Filipinos recounted to me—vividly illustrates the unwillingness of foreign workers to simply acquiesce to U.S. military contractors' exploitative working conditions and pay. Despite the risks that this entails, they have often chosen instead to fight.

Jumping

More common than protests and strikes is another type of labor activism that workers refer to as jumping. Jumping, as mentioned above, is the practice of covertly changing one's employer on a military base. The most common reason that workers jump is the promise of a larger salary at the new company.

Before going into the details of this phenomenon, it may be useful to explain why I consider jumping to be a form of labor activism. On the surface it may appear incongruous to lump it in with protests and strikes. The latter are visible (at least on military bases), mass actions aimed at subcontractors, prime contractors, and/or military officials. Their goal is to make known grievances concerning pay, food, working conditions, etc., and prod relevant authorities to remedy the situation. Jumping, in contrast, is done in secret. It is also an individual action—though one that usually requires the assistance of multiple people on a base to succeed. And it is not done with the collective aim of bettering workers' status. Put in terms of Albert Hirschman's famous *Exit, Voice, and Loyalty* treatise, protests and strikes represent "voice"—attempts by workers to repair or improve their relationship with employers through communication of grievances and desired redress—while jumping represents "exit"—an abandonment of the relationship.[13]

Why then should jumping be viewed through the lens of labor activism? The primary reason is that it represents a challenge to military contractors' power over their workforces, one perhaps even more fundamental than protests or strikes. This power derives from the fact that foreign workers on bases in warzones are a doubly captive labor pool. First, as noted above, they do not have alternative in-country work options if they are fired, rendering them more vulnerable than local laborers. Second, and more significantly, foreign laborers' right to live and work on a military base is determined by their employer. This is because they are required to have a LOA that identifies them as "contractors authorized to accompany the Force."[14] Military contractors are responsible for registering their employees with a base's "contracting officer," who then issues a LOA for each worker. In addition to specifying one's employer, LOAs also define rights and privileges on a base, such as whether or not someone is allowed to use MWR facilities, possess cell phones or computers, or move around a base without escorts. At any point in time, and for any reason, a company can inform the contracting officer that they are ending their relationship with an employee. This leads to the immediate revocation of one's LOA, followed by deportation from the base as soon as can be arranged. Right to work, in short, is entirely dependent on the whims of one's employer.

This dependency bears a striking resemblance to the *kafala*, or sponsorship, system utilized by most countries in the Middle East. As with TCNs on military bases, migrant workers' right to live and work in countries with *kafala* labor laws, including massive labor importing states like the UAE, Saudi Arabia, Qatar, and Kuwait, is controlled by their employer, or "sponsor."[15] Given this similarity, as well as the Gulf provenance of many military subcontractors, labor abuses—such as confiscation of passports, wage theft, trafficking and excessive recruiting fees, substandard living quarters, and unsafe working conditions—on U.S. bases have tended to parallel those experienced by workers in Gulf states.

Foreign workers successfully leaving their employers for better pay or working conditions with another company on a military base constitute a threat to this second aspect of labor captivity. This is especially the case when jumping becomes a widespread practice, which threatens companies' ironfisted control over their employees. That military contractors—especially subcontractors, whose workers have the greatest motivation to jump—recognize this threat is evident by the fact that their contracts often explicitly forbid employees from jumping to another company, language that is sometimes accompanied by threats of large fines or penalties. Najlaa's contracts, for example, stated that employees would be charged a $2,500 fee if they left to work for another company "before completion of one year of service." The contract Kulak has its employees sign includes a "resignation" section that stipulates, "You are not allowed to leave Kulak Cons. Co. and work on any other company." In response to an epidemic of jumping in 2003 and 2004, Serka revised its contracts with Filipino workers to include the following language forbidding workers from leaving the company: "You are not allowed to leave Serka Company and work for any other company part time or full time. If you wish to be released by Serka to another company you have to pay Serka Company the sum of **$5,000** as the transfer fee" (bold and underline in original).[16]

A second, corollary, reason that jumping should be considered an act of labor activism is that it is a strategy employed by workers to get out from under the thumb of employers, to improve their lot. Moreover, jumping carries similar risks as striking or protesting. In fact, jumping is arguably riskier as getting caught before successfully switching to another company almost always leads to termination and deportation, while not all participants in strikes and protests suffer these consequences.

Despite the risks, jumping is widespread, at least according to Filipinos I interviewed. Isko claims that other PPI employees he flew into Iraq with tried to arrange work with a new company before they even settled in. "Even though they didn't have their [ID] badges yet they were already looking to jump!" Danilo echoes this, claiming, "Most of us who could jump from PPI did." When I asked him what type of work people looked for, he replied, "Administrative work, because it pays better. And also technical work like mechanic, [on] trucks or heavy equipment." The most desirable companies to jump to, I was repeatedly told, are U.S.-based prime contractors, who pay more that subcontracting companies and offer better accommodations and base privileges. Not everyone follows this path, though. During his second "tour" of Iraq—after working for PPI for nearly three years and then heading home for six months—Sam jumped from Kulak to Jamaher Contracting Company, a Saudi firm with multiple military construction contracts. He eventually rose to the position of project manager. Sam views Kulak as "a really bad company," but one that was ultimately useful. "It was, let me just say, my stepping stone. Just my access to get back in Iraq. Because when you arrive in Iraq you have a lot of companies that will hire you."

So how does one successfully jump from one company to another? The following example is provided by Rowel. After more than two years with PPI, he decided to jump. The company he transferred to, Card Industries, provides "manpower solutions" to prime contractors in Iraq and Afghanistan, primarily in the electrical and instrumentation fields.[17] What this means is that Card does not directly hold military contracts, but recruits and "rents out" (in Rowel's words) skilled workers to prime contractors that do, such as Parsons, Louis Berger, and Bechtel. Here is Rowel's account:

> ME: Can you walk me through the process? How did you find the job? How did you negotiate [this] when you're already employed by PPI?
>
> ROWEL: By that time I heard [about opportunity] from some Filipino people, because we are friendly. We can walk outside the company [camp].
>
> ME: You'd ask around?
>
> ROWEL: Yeah, ask about other companies. They [Card] said, "Oh yeah, we're hiring some people." PPI at that time the salary is only $700 [a month]. Then if there are some rumors—actually not rumors, that's truth [about hiring]—some Filipinos start moving to the company. They got a salary like, give them like $2 grand, $3 grand [a month].
>
> ME: When you heard that [did] everyone start to think about jumping and asking [questions]?
>
> ROWEL: Yeah. Because we are here to make money. I heard the company [Card] hiring people . . . rental to another company, Parsons. [So] I applied.
>
> ME: How did you apply?
>
> ROWEL: After work, because my work is seven to five, after work I go outside, just walking around the base, because we got an ID [badge], we got access, [so] you can walk.
>
> ME: You had a green badge?
>
> ROWEL: Yes, green. Privileges. We don't need an escort [to move around the base]. We can walk in secret. Walk over to the [Card] office. Bring my ID and my passport.

ME: You didn't need a CV?

ROWEL: [I] needed a CV. But [there were] some Filipino people working in KBR office. I walked to the office, secretly talked to guys, a
friend working in the office, "Can you make my CV like this?" They
print it and give it to me after duty, 5:30. Bring CV, then if the boss,
the manager is available he can start interview, asking something
about your job and experience. If you agree about the salary they give
you, that's it. Come back [in] three days and sign your contract.

Jumping may be a strategy taken by individuals, but as Rowel's story illustrates it is also a social and collaborative endeavor, especially if one wishes to
succeed. Critical information circulates among friends and colleagues on
bases. Which company is hiring? Who is the manager to talk to? What application materials are needed? Rowel was convinced to jump to Card after working
for PPI for more than two years in part because he knew people who had already
done this and were willing to explain the process to him. Additionally, when he
needed help with an important part of the application—writing up and printing
out a CV—he had friends doing administrative work with KBR who he could
turn to.

In some cases supervisors or managers with prime contractors facilitate
jumping for those working in the skilled trades like electricians, mechanics, and
engineers, in effect poaching workers from their subcontractors. This was the
experience of Susi, a power mechanic. Though employed by PPI he worked with
KBR personnel in the Green Zone in Baghdad, installing and repairing electricity generators. After a couple of months he was approached by his boss.

The boss asked me, "Let me see your payment" [PPI's pay stub]. I gave
him the stub, and he said, "What? Is this all?" "Yes, sir. That's my salary." "It's not a salary, it's just an allowance. Do you accept this?" I said,
"Yes, there is nothing else." "No! You have no future . . . That's why we
are here, to have money, but this kind of salary is not good for you." "You
can do something about this?" He said, "Yes, I can. Wait for me. I'll just
go to the PX and talk to my friend." After a while he came back. "OK.
Tomorrow morning, first thing in the morning, 8 o'clock, we'll go to the
PX, meet my best friend." Next day in the morning, I saw the man. "Are
you ready for an interview?" "No, no, no," I said. "No need, we will go
to the workshop, I'll show you what we do" . . . Around 15 minutes
[later] he said, "Sign this contract for me. Right now. Don't think twice,
just sign it. How much do you want?" "Just give me $2,000 [a month],
sir." "What about $1,500 starting salary and then, in a few months we'll
give you $2,000?" "OK, I'll sign it then."

Susi's and Rowel's narratives may give the impression that jumping is fairly easy. But this is not the case. A variety of factors shape one's chances of pulling off a switch from one company to another. Jumping is most prevalent at large bases like KAF and Victory Base Complex in Baghdad that have thousands of workers and dozens of contracting firms, especially if such bases also have less restrictive regulations on foreign workers' freedom of movement during off hours. At small bases with fewer companies there are fewer options for jumping. These bases also facilitate closer surveillance over workers, making it more difficult to successfully arrange a transfer. As the most desirable companies to jump to are prime contractors—especially U.S. firms—English fluency or competence is another key consideration. Filipinos I talked with who jumped in Iraq and Afghanistan, such as Susi and Rowel, are generally more comfortable with English than those who did not. In this Filipinos tend to have an advantage over other TCNs from Asia. Therefore it is possible jumping may be less common overall than my discussions with Filipinos suggest.

The main reason that jumping is difficult and risky is that subcontracting companies have worked diligently to thwart the practice, often with successful results. Rene, who worked for PPI from 2004 to 2011 as an electrician, told to me he received multiple offers to jump to a new company during this time. But he never did, in part because he saw several cases where PPI convinced the military to rescind these new contracts. Confused, I asked him to explain this to me.

> ME: Let's say you successfully jumped. I offer you a job. "Come work with our company. I'll double your salary." You say, "OK." You sneak out, you come to me. Then PPI gets upset and goes to—
>
> RENE: The military. Because you [already] have a contract. [The military says to new company], "Send him back to PPI."
>
> ME: Then you get sent back to PPI?
>
> RENE: Within twenty-four hours they [PPI] will get your LOA [rescinded]. [You] go to Dubai, back to the Philippines.

Rene's account is supported by Domingo, who worked at Balad Air Base. Originally recruited by PPI, Domingo jumped to a FedEx subcontractor in 2006, increasing his monthly salary from $500 to $1,500. A year later he arranged for a friend to join him, but the move was sabotaged by PPI. "One of our friends, I helped him also. He was already hired at the FedEx [subcontractor] and then they noticed in the office in PPI. They picked him up, then they sent him back home." Workers are most vulnerable to this strategy in the two to three days it takes between signing a contract and having all the necessary paperwork—such as a new LOA and badge—processed by the contracting officer on a base. To avoid getting caught and sent home, some workers would hide until all the paperwork was complete, according to Domingo. "There was a lot of PPI workers [that] jumped but will hide first. They wait for [new] ID. If you have it [new ID and LOA] they [your old company] will not do nothing to you." Rather than retaining employees who attempt to jump, PPI and other companies pursue the punitive measure of deportation to discourage others who are looking to do the same. This is a powerful deterrent, as most people who decided against jumping cited fear of losing their job as the primary reason they made this decision.

Rescinding one's new contract is not the only strategy at companies' disposal. One of the most effective ways to prevent employees from jumping is to confiscate their passports, a widespread practice not officially banned by the military until 2006.[18] An alternative, as Mary recalls her company doing in the mid-2000s, is to impose a stricter curfew on employees. When people asked why, they were told that too many people were jumping. The curfew was accompanied by increased surveillance of workers, including bunk checks (to identify anyone hiding while waiting for paperwork to be finalized) and searching bags when people left their camp in the morning. The reason for this latter check, according to Mary, was that workers wanted to get their clothes and possessions outside the camp before jumping and so would "bring their clothes little by little in a bag. Then they don't have to come back anymore if they already got their clothes." If caught with clothes you could try to fool the guards by claiming you worked in laundry operations and saying, "I'm not jumping. I'm just doing my laundry . . . you know [at] the big American laundry" because people who worked there often did this due to the higher quality of washing machines compared to those provided in company camps. Finally, several people who worked in Balad told me that sometime in 2007 or 2008 the main military contractors on the base reached an agreement not to steal each other's employees, an action not dissimilar to the informal anti-poaching accord reached by Apple, Google, Intel, and other firms during the same point in time in Silicon Valley. But unlike the plight of Silicon Valley engineers, this instance of anti-labor collusion was never covered by the media, or the subject of a federal, class-action, anti-trust lawsuit resulting in a multimillion dollar settlement.[19]

The lengths to which companies have been willing to go to stem the tide of jumping extends beyond military bases where it takes place. After a rash of workers quit the company in 2004 and 2005, Serka began putting pressure on its recruiting network in the Philippines. Michelle, the local agent responsible for recruiting dozens of Serka food service workers, recalls that Serka "called the [recruiting] agency" and then the "the agency called people here to inform the wives that their husbands jumped," which they were reminded is a violation of their contracts. Another recruiting agency remembers the subcontractor they worked with threatening to withhold payments if they did not do a better job of screening potential applicants and dissuading them from jumping. The agency's manager replied, "Why do you put all the fault on me? Everyone says 'Yes, ma'am.' The problem is that when they get there [Iraq] they hear that the salary that you're giving is not enough." PPI went even further, instructing its Philippine recruiting agency, AES, to file a "complaint for disciplinary action for breach of contract" with POEA against at least nineteen former employees in 2005 and 2006. These complaints argued that by jumping these individuals caused "sustained damages" for PPI and AES, and "tarnished" and "besmirched" the reputation of the country's recruiting industry. In two cases POEA ruled in favor of the companies, temporarily suspending the rights of the former employees to work abroad for a period of two to four months.[20]

These legal proceedings demonstrate again the argument I made at the beginning of the book that one of the main effects of military logistics contracting is the generation of various entanglements—economic, social, political, and, in this case, legal—which extend well beyond the battlefields in Afghanistan and the Middle East. We will return to this point later in chapter 10, which examines the themes of family, community, and returning home. But first I want to explore further life, work, and social relations on bases.

RELATIONS

> I saw it as a caste system. It was Americans on top, then the
> Europeans underneath, then Filipinos, then Indians. . . . But that kind
> of dividing of people is just wrong for everybody. What's the differ-
> ence between me and the black American guy, or me and that TCN?
> What's the difference? We all did the same jobs. We were all in the
> same base.
>
> —Lena

It is difficult to decide where to start, or how to organize, a chapter that aims to
describe social relations on military bases in the Middle East, Afghanistan, and
Africa, given the diversity of experiences and settings. Any account, as noted in
the introduction, will be partial and incomplete. That said, there are certain themes
that stand out based on interviews. This chapter examines two social fields that
significantly influence the experiences of the military's TCN workforce, which in
their most basic form can be referred to as company and identity, respectively.
To be certain, these are inextricably entangled. Thus this is a somewhat artificial
distinction that I am making, as will be clear in the analysis below, which also
emphasizes intersections and connections among them.

Perhaps the most important influence on life on a military base in a warzone
concerns the type of company that one works for. The key distinction is that between prime contractors—especially U.S. firms—and their subcontractors. It is
hard to overstate just how important employment with a prime contractor or
subcontractor is in determining pay and privileges, as well as relations among
workers, and between workers and service members. Bosnians employed by
prime contractors, for instance, have typically lived in housing with American
workers, enjoying similar base privileges and competing for jobs and promotions. Their lives are a world apart from other TCNs working for subcontractors.
Another way to illustrate the prime contractor-subcontractor distinction is to examine accounts of individuals from the Philippines who have managed to jump
from a subcontractor like PPI to a prime contracting company. I also discuss in
further detail below the introduction of tiered contracts by Fluor and DynCorp

in Afghanistan, which has blurred the distinction between prime contractors and subcontractors in recent years.

The second sphere, identity, is multifaceted. For instance, patterns of recruitment by prime contractors and subcontractors—with the former primarily hiring workers from the U.S. or Southeast European countries and the latter sourcing labor predominately from South and Southeast Asia—highlight the role that differential pay and privileges play in contributing to racial disparities within the military's workforce. Indeed, several researchers have argued that the extensive recruitment of workers from countries in the Global South reinforces racialized hierarchies in warzones. The above observation from Lena, who worked for multiple companies in Kuwait, Iraq, and Afghanistan, points to another significant aspect of relations on bases: one's citizenship or nationality. As her quote indicates, this also intersects with race, both within—such as American—and across national categories. Possibly the least explored aspect of base life concerns the experiences of female workers like Lena, as with few exceptions news accounts and academic articles overlook the fact that a not insignificant number of contractors are women. Race, citizenship, and gender often intersect in complex and unexpected ways, which I discuss in the second section of this chapter, following an examination of the divergent experiences of prime contractor and subcontractor employees.

Company

Most accounts of foreign labor on bases in the Middle East and Afghanistan highlight the exploitation of this workforce by military contractors, from low pay and poor living conditions to trafficking. In nearly every instance the offending companies have been subcontractors, predominately from the Gulf states and Turkey. Often unseen in these accounts is the experience of those who work for prime contractors. The differences are stark. Consider the following exchange with Fedja, who was employed as a labor foreman by KBR at Tallil Air Base in Iraq in 2007:

> FEDJA: You know in Iraq I had a CAC card from the U.S. government—a white CAC card—I had every rights of an American citizen [on the base].
>
> ME: Is that because you had previous experience working with American forces here [in Bosnia]?
>
> FEDJA: No. It was the KBR contract in 2007. And in those days we had the white CAC cards.
>
> ME: What is a white CAC card?

FEDJA: It is like this. You had a couple of cards. CAC card is your chip card, you had everything in it—with the CAC card you go on R & R [vacation], you sign in to your outpost—everything is on your CAC card. And we had a white CAC card like American citizens [contractors].

ME: So these are electronic?

FEDJA: And visual. Because in every base you have inside security. It was visual security for inside the post because you had limitations. For my base, for example, you had people from Turkey, from India, from all around the world. And they could go to work, and then back into their outpost [mancamp]. Because inside a base you had ten other bases [company-run mancamps]. I, with a white CAC card, I could go to PX, to a gym. I could shop and buy everything I want. I could go to pizzeria. But people who didn't have a white card couldn't go. They could only go to their base [mancamp]. They had their own mess hall and everything.

ME: So you had a segregated mess hall then? What different cards were there? Or what different levels—let's say for different workers—were there?

FEDJA: It was like two types of levels. Minor jobs like cleaning the toilets, those really, really low jobs were being done by people from India, and they had major restrictions. They could only go supervised to work their job. I picked them up from their base [mancamp], inside this base they are quarantined, a small base for them only.

ME: And they couldn't leave it unless you came and supervised them?

FEDJA: I would come and pick them up, do my eight hour shift, drive them back and that's it. They can't go to mess hall with U.S. troops. They can't go anywhere without supervision.

ME: And your job was?

FEDJA: I was a labor foreman. I supervised people who worked for me.

ME: How many people would you have, then, as a foreman?

FEDJA: From three to eight guys.

ME: And they would be Indian or Pakistani?

FEDJA: Yeah.

ME: And what were they doing?

FEDJA: Most of the time they were doing the cleaning jobs. Latrines, showers, and stuff like that. That's all the jobs they could do because the contract—KBR had a lot of jobs—and all the important jobs were you needed experienced people, like air conditioners, electricity,

> power generators, stuff like that, they were hiring Bosnians for an
> excellent salary. And third country nationals they were doing the
> lowest jobs, cleaning, nothing else.

Fedja's comments highlight several of the most important contrasts created by
the prime contractor-subcontractor system. First, as a KBR employee he had many
of the same privileges accorded to U.S. contractors and soldiers at Tallil. He could
eat in military DFACs, use MWR facilities, buy sundries at PX stores, and move
around the base with few restrictions. This status was exemplified by his white
CAC card. These identification cards are issued to all active duty uniformed personnel, DoD civilian employees, and select military contractors. At Tallil KBR employees like Fedja received CAC cards, which provided visual indication of their
privileged status. In contrast, employees of subcontracting companies at the base,
the largest of which in 2007 were Kulak, GCC, and Iraq Projects Business Development, received color-coded "badges." Though these badges have varied across
bases and the period when they are issued, they typically provide an employee's
name, company, and an identification or passport number. Badge colors indicate degrees of mobility and access to facilities at a base. The Indian and Pakistani workers that Fedja escorted wore red badges, which meant that they were
forbidden from moving anywhere on the base—except inside their company-run
mancamp—without an authorized escort. In Iraq badges usually ranged from red,
orange, and yellow to green or blue, with the latter colors indicating that the wearer
was allowed to move around a base without escorts, and even had access to certain facilities like PX stores.

A second difference between prime contractor and subcontractor employees that this conversation raises revolves around work, pay, and contracts. The
men Fedja oversaw were tasked with "minor jobs"—cleaning latrines and
showers—in his view. For the most part subcontracted labor at bases in the
Middle East, Afghanistan, and Africa is used to perform similar low-skilled or
poorly paid work such as DFAC operations, laundry services, cleaning, construction, and basic warehouse tasks. Most Bosnians who have worked for
KBR, Fluor, or DynCorp, in contrast, have been hired for skilled labor positions such as electricians, mechanics, or heating, ventilation, and air conditioning installers, have performed administrative tasks like property management
or payroll, or have held supervisory positions like QA/QC inspectors and labor
foremen. These jobs tend to pay extremely well, especially compared to salaries
earned by subcontractor employees. As a labor foreman, for example, Fedja
earned $5,300 a month, more than ten times the salary of his Indian and Pakistani
charges. This despite the fact that his job actually had few particular skill
requirements.

Typically, even for similar jobs there are considerable differences in pay depending on whether one works for a prime contractor or subcontractor. Several Bosnians I interviewed had worked in material management and supply operations, which involves moving, sorting, and tracking military and contractor materials in warehouses. These are well-paid jobs, ranging from the high $30,000s to $70,000 a year for those who worked for KBR in Iraq or on European contracts with Fluor and DynCorp in Afghanistan. Filipinos who perform similar tasks at bases in those countries but are employed by subcontractors like PPI are paid but a fraction of this amount. In addition to the pay differential, as a KBR employee Fedja was given three paid leaves ("R & R") a year, while the men he supervised received no leave during their two-year contracts.

A third point concerns differential relations between American employees and foreign labor depending on whether one works for a prime contractor or not. In the above passage Fedja refers twice to the fact that his "white CAC card" was the same as those issued to Americans, and carried with it the same privileges. In contrast, the subcontractor employees he supervised were confined to their camps when not performing the menial jobs that were reserved for them. Moreover, his characterization of them as a category apart ("third country nationals") is remarkable because, as a Bosnian national, Fedja was of course himself classified as a TCN by the military. Yet, as discussed in chapter 4, he and other Bosnians who worked for KBR under the LOGCAP III contract did not see themselves as such, in large part because the company treated all of its employees (American and foreign citizens alike) as part of a "KBR family" that stood apart from and above those working for its subcontractors. The following exchange about base housing arrangements is illustrative.

> ME: Were you mixed in with soldiers?
> FEDJA: No. We had different [housing]. We called them hooches. The Army was separated. KBR was separated. Third country nationals were separated.
> ME: So you weren't called a third country national?
> FEDJA: That's the line that I picked up from the Americans [KBR employees], third country nationals. Because they were looking at them [Asian workers employed by subcontractors] as third country nationals.
> ME: But they weren't looking at you as one?
> FEDJA: No, because they worked with me and we had the same CAC card and the same paycheck.

As I discuss below, this does not mean that there were no tensions between KBR's Bosnian and American workers, just that for the former, such as Fedja, the relevant "comparables" or "comps"—to borrow a term from the real estate industry—

were American contractors, not other foreign workers on the bases. I should also note here that KBR is somewhat of an outlier in the efforts to which it goes to inculcate a distinct corporate identity with its employees. This said, when it comes to the chasm separating its employees and other prime contractor workers from subcontracted labor concerning pay, privileges, and status, the difference is a matter of degree not kind.

This is perhaps best illustrated by describing the experiences of Filipinos who successfully jumped from subcontracting companies to jobs with prime contractors. Take Rowel, whom we met in chapter 8, who worked as electrician for PPI for two years in Iraq before jumping to Card Industries, which "rented" him to the giant U.S. engineering firm Parsons. After this contract ended in 2009, he worked for another U.S. engineering company, Louis Berger, in Afghanistan. Like all Filipinos I talked with who jumped, his primary motivation was money, in his case an increase in salary from $700/month with PPI to $2,300/month with Card. But in our conversation he also emphasized the difference in privileges, such as use of cell phones and computers to keep in touch with his family:

> In PPI you don't have computer. You don't have cell phone. You got only like five minutes a week. In a week only five minutes privilege on a [company] cell phone. Sometimes they are busy. There's always [a] low bat[tery]. Your five-minute free time [you] keep calling them, they are busy. The phone is busy. After that one you cannot talk to them. You have to wait a week again to talk to them. Not like in Card and Louis Berger, [where] you got access. Every day you got phone. . . . Every day we talk because I got [my] own computer on my job because I work in a power plant. We got [our] own office . . . After [work], I go to the computer. If they are online, [I] talk to them.

A second difference Rowel stressed was living conditions. At the PPI man-camp in Victory Base Complex there were ten men living in a forty-foot shipping container. The company's camp also had only ten showers for hundreds of people, so you had to get up early in the morning, because "if you're getting up late, no hot water." In contrast, at Card he shared a twenty-foot container, which had its own bathroom, with one coworker. While with Card he also received a $300/month cash allowance—nearly half his previous salary—which could either be saved, or "if you don't want to eat in military DFAC, go to Popeyes [Chicken], go to the coffee shop [Green Bean], drink coffee with those muffins."

I then asked Rowel if his American managers or coworkers at Parsons or Louis Berger used the term TCN when referring to his status. He said, "Yes," which was followed by this exchange:

ME: What does that mean? What do you think about that?

ROWEL: I'm not thinking hard about that one because we are Filipino. The meaning of TCN for me is third country nations. It's OK for me. Because when I work in Louis Berger, like I said they gave me full of access that's why I'm not thinking about I am a TCN. The military guy go to MWR, doing gym, I can go there also, same what they did. See what they're doing, I can go [on] R & R. I can watch [TV] on the MWR. Whatever they eat, I can eat also.

Like Fedja, this term had little meaning for Rowel as a prime contractor employee because it did not reflect the privileges, pay, and status he was afforded compared to his previous experience with PPI.

Rowel's perspective is echoed by others who jumped. After years with PPI in Iraq and Afghanistan, Fidel landed a position with KBR in Africa, along with several other workers from the Philippines. He described the difference this way: "In Iraq and Afghanistan we [he and his fellow Filipinos] are TCN. In Africa we are expats. They treat us as expat. Our accommodation is good. We have one [an individual] room, air conditioning, one bed. We have cable TV. . . . Every morning there's a local [that] pick[s] up our laundry. Then in afternoon return[s] it." Susi, who jumped to Arkel in Iraq, worked on DoD-funded civil power generation reconstruction projects. As this job required travel to various sites across the country he was given a CAC card granting him access to amenities on all bases in Iraq, and permission to carry a service pistol, radio, and telephone. Compared to PPI, with its "limited" privileges, as an Arkel employee he was afforded, in his words, "full access." When asked if this meant he was considered a TCN when working for Arkel he replied, "No," and then, "That's why I'm telling you we are not all the same experience."

"I'm telling you we are not all the same experience." Susi's admonition is worth keeping in mind when discussing the military's foreign workforce. It even applies to the prime contractor-subcontractor divide, perhaps the single most determinative factor shaping work and life on overseas bases in warzones. The best example of this is the introduction of tiered contracts by Fluor and DynCorp in Afghanistan in 2010, which has led to a relative blurring of lines. As discussed in chapter 4, these contracts divide the companies' employees into multiple tiers based on nationality and geography. DynCorp's categories are Expats (Americans), Foreign National United Kingdom (FNUK), Foreign National European (FNE), and Foreign National Asian (FNA). Fluor has set up a five-level classification system that distinguishes between company staff, Americans hired on contract, West European employees, East Europeans, and workers from Asia. The primary impetus for the introduction of these tiers appears to have come from the Pentagon, which directed LOGCAP IV prime contractors to bring salaries for foreign direct hires more in line with prevailing wages in their home countries. Thus, starting in late 2008, KBR also lowered its pay scale for new Bosnian recruits to Iraq and Kuwait. But due to the drawdown of forces in those countries, this change has primarily affected those recruited by Fluor and DynCorp, who took over LOGCAP operations in Afghanistan. At the peak of operations in 2011 the two companies and their subcontractors provided support for 133 bases and nearly 100,000 U.S. troops in the country.[1]

Fluor's and DynCorp's foreign employees—particularly those from the Balkans—experienced a reduction in both pay and status under this tiered system. Bosnians hired by Fluor, for example, are paid 45 percent of what Americans and West Europeans earn for the same jobs, with a similar gap in pay between DynCorp's workers from the Balkans hired on a European contract and American labor. Those who have an Asian contract with DynCorp are paid even less, typically earning $12,000 to $18,000 a year—roughly one-third the amount paid to those doing similar jobs under a European contract. A commensurate shift has occurred when it comes to status. Whereas KBR developed a culture that treated its employees (both American and TCNs) as a company family positioned above its subcontractor workforce—and continued to do so under LOGCAP IV according to Bosnians who have worked for the company during this period—those working for Fluor and DynCorp report that they are frequently reminded of their lower status. This ranges from the common use of terms like TCN or OCN in conversations with American coworkers to little details like Damir's example in chapter 4 of the use of separate buses for American and foreign workers upon arrival in Bagram.

Interviews in Bosnia and the Philippines suggest that the introduction of tiers has had the greatest impact on those who work for DynCorp, due to the fact that the company has direct hired significantly more workers under Asian contracts than Fluor, which has followed KBR's practice of relying primarily on subcontractors for low-skilled and low-paid labor. One reason for this difference is that DynCorp found itself shorthanded in late 2009 when one of its two primary first-tier subcontractors, Agility (formerly known as PWC), was barred from receiving government contracting money following a lawsuit accusing it of overcharging the military billions of dollars under its DLA contract to provide food for troops in Iraq and Kuwait.[2] Short subcontracting support, and behind on several projects, which invited criticism from the military and government auditors, it appears DynCorp executives in Afghanistan decided to turn the company into its own body shop by direct hiring thousands of workers from Asia, Southeast Europe, and Africa (primarily Kenya) under "Asian" contracts that offered pay and benefits similar to its subcontractors. According to Diana—who states she "didn't speak no English at all" at the time—DynCorp's recruiting process for Asian contracts in late 2010 more resembled PPI's early scramble to hire workers from the Philippines than KBR's and Fluor's more exacting standards: "We went over there in Hotel Tuzla and they did some kind of interview but . . . they didn't ask us a lot of questions. Of course we already know what they will ask us, let's say, 'What is your first name, last name?' Because we will work in laundry. We don't need that much English. We will wash, clean and that's it. . . . After five, ten days, they call us that they will hire us and next month that we will have a flight to Dubai." Her salary was a $1,000 a month. Sead, a young bartender from Tuzla who was hired as warehouseman under an Asian contract around the same time as Diana, recalls that his group had "guys from Kenya, from India, most of them. And you had Filipino guys, and Bosnian guys." Grace, a Filipina who worked for DynCorp for two years in Afghanistan on an Asian contract that paid $1,400 a month, remembers sharing a large Alaska tent with other female DynCorp direct hires at Camp Dwyer: "We had Americans there. Kenyans, Macedonians, Bosnians."

Identity

Fluor's and DynCorp's introduction of a tiered pay scale highlights the degree to which the military's logistics workforce is stratified along racial and national lines. Though the number of tiers and labels differs across the two companies, they both essentially divide their workers into four hierarchically ordered categories: Americans, West Europeans/UK citizens, Southeast Europeans, and Asians. The greatest difference in pay is that between those with Asian contracts and the rest. In this the companies' internal tiers mirror a broader racial hierarchy on military bases in warzones. Painting with a broad brush—and stressing that this is an oversimplification with numerous exceptions—the primary distinction is between relatively high status and well-paid American and European workers, and a poor, often exploited, Asian workforce.

Nearly every journalist and scholar who writes about foreign labor in the wars in Iraq and Afghanistan has highlighted these disparities, both in the fields of logistics and security. Indeed, the racialization of labor may be even more prevalent in the private security industry, where discourses extolling former colonized peoples from the Global South as "martial races"—such as Gurkhas and Fijians—abound.[3] None of this is new. Indeed, when it comes to the experiences of logistics workers there exists a remarkable parallel from a century earlier: the "silver and gold" system set up by the U.S. during the construction of the Panama Canal.

The Panama Canal is an engineering marvel, celebrated as the "eighth wonder of the world" upon its completion in 1914. But it is also as much a feat of labor as it is engineering. In fact, in 1906 the project's chief engineer, John Stevens, claimed that "the greatest problem in building a canal of any type on the isthmus . . . is the one of labor. The engineering and constructional difficulties melt into insignificance compared to labor."[4] To surmount labor challenges U.S. administrators recruited widely, bringing in tens of thousands of workers from across the Caribbean, Central America, the U.S., and Europe. They even proposed recruiting Chinese laborers, but this scheme was rejected by then-U.S. attorney general William Moody, who argued that the importing of "Oriental aliens" under contracts to perform labor "is not necessarily one of involuntary servitude, but it may be and, in fact, usually is a condition of involuntary servitude."[5] A century later the U.S. now looks the other way as tens of thousands of South Asians on its bases in the Middle East often work under contracts and conditions of debt bondage that also constitute involuntary servitude.

To manage its massive and diverse workforce in the Canal Zone, U.S. administrators established a segregated silver and gold system. Under this system, "the government paid silver employees far less, fed them unappetizing food, and housed them in substandard shacks. Gold workers earned very high wages and terrific benefits, including six weeks of paid vacation leave every year, one month of paid sick leave every year, and a free pass for travel within the [Canal] Zone once each month."[6] Like the prime contractor-subcontractor system today, the silver and gold system was largely, but not exclusively, organized around racial distinctions, though it began as a more fluid way to reward productive employees regardless of race or nationality. In 1906 Stevens issued an order requiring "colored employees" from places such as the West Indies to be placed on silver rolls while white Americans were placed on gold. All but a handful of African American workers who had been explicitly hired on gold roll contracts were also shifted to the silver roll. Somewhat paradoxically, the silver and gold system also revolved around citizenship, especially following an executive order by President Theodore Roosevelt in 1908 that stated that gold roll employment should be limited to U.S. citizens. This resulted in the shift of a number of European laborers from gold to silver rolls. At the same time the U.S. decided that Puerto Rican and Panamanian workers should be eligible for gold roll employment due to the former's status as colonized "wards of the nation" and the latter's position as citizens of the country in which the canal was being built.[7]

There are other parallels between the silver and gold system and present-day military labor practices produced by the prime contractor-subcontractor system. In addition to pay, both formed the basis of the distinction between supervisory and supervised work. For example, in Panama "one gold carpenter (typically a white U.S. citizen) might oversee eight to twelve silver carpenters (West Indians); one gold plumber might manage an area with a few silver plumbers under him."[8]

And in both contexts a divide and conquer strategy was used as a means of facilitating a more docile workforce. In 1906 the chairman of the Isthmian Canal Commission, the body originally charged with overseeing construction of the canal, claimed that "a labor force composed of different races and nationalities would minimize, if it did not positively prevent, any possible combination of the entire labor force."[9] Despite this, strikes over food and wages in Panama were not uncommon. So too was the strategy of moving from lower- to higher-paying jobs by silver roll workers, though the more rigid delineation of these categories along racial and national lines limited the ability to substantially improve one's station through this strategy compared to Filipinos and other subcontractor employees who jump to positions with prime contractors.

One difference between the two systems, as detailed in the previous two parts of the book, is that the racialized hierarchy of labor at bases in the Middle East and Afghanistan is less a product of intentional policy by the U.S. government—as it was in Panama—than the intersection of historical circumstances with contrasting recruiting patterns and labor practices by prime contractors and subcontractors. That KBR's direct hire TCN workforce in Iraq was overwhelmingly from Southeast Europe was directly related, for instance, to the fact that in the late 1990s and early 2000s it hired tens of thousands of people from Bosnia, Kosovo, and Macedonia as LN labor when it provided LOGCAP support for the peacebuilding missions in the Balkans. At the same time that these missions were beginning to wind down in the mid-2000s, U.S. military activities in the Middle East were ramping up, thus many were subsequently employed by KBR when it found itself shorthanded in the early years of the occupation of Iraq. Once this recruiting pattern was established, it was logical for Fluor and DynCorp to also turn to the region to fulfill direct hire labor needs, especially considering that they took on much of KBR's workforce in Afghanistan following the transfer of LOGCAP support for that country to their hands in 2009. Similarly, the prevalence of workers from South and Southeast Asian countries is connected to the provenance of military subcontractors in CENTCOM, most of whom hail from the Persian Gulf or Turkey. When these firms utilize recruiting agencies in countries like India, the Philippines, Nepal, and Sri Lanka to amass the pool of labor needed to fulfill their contractual obligations in Iraq, Afghanistan, and other countries in the Middle East, they are drawing on well-worn pathways that constitute a massive labor import-export regime between wealthy Gulf petro-states and poor Asian labor-exporting countries. At the same time they have brought with them exploitative labor practices that characterize operations in home countries.

The racial disparities that exist on U.S. bases in CENTCOM, in other words, are more a product of contingency than intentional design by military officials. Nonetheless, through its actions the military has been complicit in perpetuating and even deepening inequalities, from its refusal to substantively combat trafficking by subcontractors to instructions to prime contractors to introduce steeper pay differentials for direct hires from Asia and Southeast Europe compared to American and West European employees under LOGCAP IV.

Compared to race, relatively less attention has been given to the role that citizenship and nationality play in structuring experiences on military bases in warzones, and the ways that the latter intersect with the former. Yet as with the silver and gold system in Panama a century ago, present-day disparities in pay, privileges, and risk cannot simply be reduced to race. When it comes to categorizing its workforce, for instance, the fundamental distinction made by the military is the line dividing American citizens on one hand, and foreign workers, both TCNs and LNs, on the other. As discussed above, on a daily basis this distinction carries greater weight for subcontractor employees than those who work for prime contractors and thus have privileges that are more comparable to Americans. But the distinction is still ever present—and it can crystallize at a moment's notice, to significant effect.

One example that illustrates this point was the military's response to the Chelsea Manning leaks in 2010. Srdjan, who was working as a logistics coordinator for KBR at Balad at the time, remembers that following the leaks every non-American worker on the base was immediately viewed as a security threat, despite the fact that the information had been leaked by an American soldier.

> SRDJAN: After the Bradley Manning case and shit we started to be treated like fucking spies. I still have all those emails. No electronic devices whatsoever. No laptops. No freaking cell phones. Nothing.
>
> ME: Did they do that with soldiers as well?
>
> SRDJAN: No, no, no, just foreigners. We were so pissed off. It came to the point that we had a meeting, an all hands meeting [of KBR employees]. "Guys, your Motorolas? Go back to your rooms and turn them in." How the fuck are we supposed to work without a radio? It was so fucked up. I had to literally—my only lifeline back home was Skype. I had bought a laptop, external antenna, got Wi-Fi from local provider that was an arm and a leg per month.
>
> ME: From an Iraqi provider?
>
> SRDJAN: Yes. And if you don't get rid of that stuff yesterday you can lose your job, get prosecuted, blah, blah, blah. I literally had to say, "Here's my laptop" [give it up]. I was pissed! Right after that meeting expats [American KBR employees] had their [own] meeting and they were told not to help us, because they could lose their jobs. So I couldn't go to my buddy, Alan and say, "Alan, please help me and hold onto

my laptop until I'm on R & R and can take it home." No. if he's found with two laptops in his quarters he would get in trouble. Our friends were pissed about it, U.S. guys and the military too. So there was an officer who said, "Srdjan, use my computer." It was fucked up.

ME: So was it all foreign nationals or TCNs?

SRDJAN: Everybody. British—non-U.S. citizens. Period. Ridiculous. And then you figure out that's the world of the military, you know? This might sound ugly, but there is no military intelligence . . . it was hard at that time. I know a guy from my hometown got fired because of a memory stick in his cargo pants. It was used for training new employees on how to load up cargo planes, it had pictures on it. Our [KBR] managers, U.S. guys, confirmed to the military, "Yes this our employee, he is an instructor. He needs this for training."

It is necessary to note that any contract laborer—American and foreign alike—can be immediately fired and removed from a base for breaking military rules. But as this example shows, citizenship is central to the military's calculations of security risk, and thus when it comes to surveillance, job security, and the extension and removal of privileges such as possession of computers and cell phones, non-Americans' positions are always more precarious and contingent.

Another issue raised by several Bosnians who have worked for KBR, Fluor, or DynCorp concerns slights by American coworkers, especially African Americans. This is alluded to in Lena's comment that begins this chapter, when she asks, "What's the difference between me and the black American guy, or me and that TCN?" Asim provides an example from his time working for Fluor in Afghanistan, involving his supervisor, Alonzo:

One day [he] approached us . . . and he said, "Hey guys I don't want to hear Bosnian any more over here." I told him. "It's my right, my human right, to speak my language with my people. Do you understand how stupid it is to speak English with this guy? Of course you do not speak my language, so I will speak English with you. But that man is Bosnian, I cannot express myself with English as well as I can with Bosnian." And he was reported to the site manager [who] said, "Stop doing that shit to people. You don't have the right to do that. I will report you next time to HR [human resources, which deals with discrimination claims]." And from that date he [Alonzo] hated everyone that come from Bosnia. Small, big, he hated Bosnians.

When I asked Goran about tensions between American and Bosnian workers at KBR, he admitted they existed but suggested this was to be expected, and that in some cases the problem lay with his Bosnian compatriots, especially when it came to interactions with African Americans.

> That's normal. It's not just Americans . . . First, we're different cultures. We see things differently, 10,000 kilometers between [the] two of us. We are Westernized, but we're different cultures. . . . I had my share of issues with some of the people, but nothing really much. I mean, that's a normal thing. Because after all, we were the outsiders. We were outsiders, and if you cannot deal with that, I mean, what the fuck? Most of the people from here, they didn't see a black guy before. This is a country where black people, they don't live here. And it was cultural shock for some of our guys to go over there and interact with different races if they didn't work here [in Bosnia] for Brown & Root.[10]

Most Bosnians, however, suggested that tensions with American coworkers were rooted in their subordinate status as TCNs. Faruk, for instance, explicitly linked what he saw as mistreatment by African Americans to racial inequality in America. "It's a power trip. It's the only time in their life when they are being above somebody else. Just because of the nationality. And so they were abusing it in the worst possible way." While this claim is impossible to substantiate, given the broad pattern of racialized labor inequality on military bases in Afghanistan and the Middle East, it is understandable that African American contractors might be especially concerned with policing status hierarchies based on citizenship.

It would be a mistake, however, to suggest that tensions between American and non-American employees working for prime contractors revolve primarily along racial lines, even if in interviews several Bosnians highlighted such cases. For example, if one peruses English-language internet job boards where LOG-CAP opportunities are discussed (such as indeed.com), along with other prominent online fora for military contracting information and conversations like blogs and Facebook groups, it is not difficult to discern a persistent line of Trumpian "America First" resentment among American workers that prime contractor jobs are being given to non-Americans, especially people from the Balkans. One site where this viewpoint was often expressed was mssparky.com, a blog run by former KBR electrician Debbie Crawford (a white woman from Oregon) from 2008 to 2013. Crawford's posts about financial malfeasance and shoddy construction work by contractors like KBR—peppered with leaked documents from a network of sympathizers working for contractors in the Middle East and Afghanistan—as well as discussions about job opportunities quickly made her blog a must-read for those concerned with military contracting. By early 2010 her site was receiving nearly 2 million page hits a month.[11]

Comments on an October 2009 post discussing Fluor's plans for transitioning over KBR employees in Northern Afghanistan under the new LOGCAP IV contract give a sense of the anger and resentment directed toward Bosnians and other TCNs. I highlight here just two comment threads from that post—which received more than 300 comments in total (all posts have been copied as in original thread).[12] The first, raised by someone that went by the moniker "Speicher Dude" (suggesting he worked at Camp Speicher in Iraq), highlighted recruiting efforts by Fluor in the Balkans, prompting a critique by Crawford of the U.S. government's refusal to prioritize hiring Americans for these jobs, and a response by another U.S. commenter who went by the name "Gijane," who suggested that maybe he should pretend to be from the Balkans.

> SPEICHER DUDE SAYS: *February 13th 2010 at 6:07 A.M.*
> Fluor is currently holding job fairs in the Balkans for future Afghaniland
>   employees . . . Soooo, if you're thinking about how many TCN's are
>   currently on LCIII [LOGCAP III]; wait till you see whats in store for
>   LCIV [LOGCAP IV].
> GIJANE SAYS: *February 27th 2010 at 7:18 P.M.*
> I understand that it is cost efficient for Fluor to hire Balkan personnel
>   (no offense) but what about the people who have more credentials
>   and experience than those people?
> Ms SPARKY SAYS: *February 27th 2010 at 7:21 P.M.*
> That is a valid point. But regardless of whether they are more or less qual-
>   ified. The DoD should hire the people who will be filling their bud-
>   gets with US tax dollars. The Bosnians or any other TCN won't! HIRE
>   AMERICANS FIRST!
> GIJANE SAYS: *February 27th 2010 at 7:34 P.M.*
> MsSparky, I know me and among hundreds of other Americans are try-
>   ing to wait for "the call" while recruiters goes to other country and
>   give away positions like candies. I think it is far double standards. So
>   what does it takes for people like me, other than a great resume with
>   not just "bullets" of duties performed but with "achievements" to get
>   noticed by recruiters? Perhaps, changing my name into Balkan would
>   kick it up a notch.

Later that year there was a much more vitriolic exchange between a commenter who went by the name "FN" (for "foreign national"), Crawford, and "Eric," another reader from the U.S. It began when FN defended the hiring of foreign workers and asked people to "keep politics out" of the discussion. The following comments ensued:

Eric says: *November 19th 2010 at 6:08 P.M.*

FN

The only good FN's are the Brits other then that they are trash taking are taxpayer dollars. You do not see any Philipino soldiers fighting this war, you do not see any Bosnians fighting this war, You do not see any Indian soldiers fighting this war, I will come out very clearly they are worthless bloodsucking leaches living off the American tax system and these companys along with the US governement should be ashamed of themselves.

FN says: *November 21st 2010 at 5:43 A.M.*

Eric

Well man, I'm sorry you feel that way because just the same as you everybody else is trying to make a living.

And I'm not talking about who is fighting the war, look hats off to the soldiers doing their job man, they are heroes. I'm talking about the people who is actually working for companies as a contractor such as KBR, FLUOR, DYNCORP. And just to notify you, mostly all of the FN's does have taxes to pay when they get back home it's just not as much as the American tax system. Maybe some of the FN's you've met or came accross are trash well let me tell you not all of them are the same and believe me I've met some of those FN's even from my own country. I've been givin more recommendation letters by the US companies than I have certificates so never judge people in quantity because you don't know all of them.

Eric says: *November 21st 2010 at 6:41 A.M.*

You must be from the Balkans Im guessing FN, some of these guys are great people but I am still paying for there salarys with my taxes. Your taxes do not go back to the United States the country who is paying for this war. Your taxes go back to your country that has not sent a dime over here. Besides I have not found one Balkan who can bend a piece of Conduit or even make job look half way presentable. YOU ARE NOT A QUALIFIED ELECTRICIAN IF YOU DO NOT KNOW HOW TO USE A FREAKING PIPE BENDER.

Ms Sparky says: *November 21st 2010 at 3:29 P.M.*

I agree. Americans are paying for this war and should have the first shot at the jobs. I don't even want to hear about how much cheaper FN's are to hire. The DoD has proven time and time again they could care less about cost savings. Allowing the contractors to hire FN's especially when they use labor brokers ehuman trafficking and abuses.

> And . . . from an electricians point of view, unless you have been trained to the National Electrical Code and certified or licensed in the States then you are not equal to an American electrician to work on a US military facility that requires work to be done to the National Electrical Code.

These exchanges present remarkably ugly and resentful comments directed toward a foreign worker for taking an "American" job—following the dubious logic that as taxpayers Americans should have priority for this type of overseas military work. Bosnians and other workers from Southeast Europe bore the brunt of these remarks, which makes sense since unlike South and Southeast Asians working for subcontractors they were frequently in direct competition for jobs and promotions with American coworkers at prime contractors. It is also not inconsequential that the Great Recession in America hit industries like construction and manufacturing especially hard, thus for some blue collar Americans military work in Afghanistan and the Middle East represented—as it did for people from the Tuzla region—an answer to economic precarity. Undoubtedly, similar sentiments to those expressed by Crawford, "Eric," and "Gijane" were held by Americans working on bases across CENTCOM, though they likely would not have been expressed as openly due to the fact that this could lead to warnings and even sanctions from supervisors and human resources administrators.

The experiences of Filipino workers also illustrate the ways in which nationality complicates narratives that emphasize a rigid racialized hierarchy of military labor in overseas warzones. Most Filipinos I talked with argued that they occupied a relatively privileged place on bases—at least compared to other subcontractor employees. One reason for this is the presence of Filipino-American personnel in the armed forces. Several workers brought up their connection with Filipino-American "brothers" in the military during interviews. For some this was primarily a social relationship, such as attending church together on Sunday, or playing pickup basketball at the MWR during off hours. But Filipino-American soldiers also served as sounding boards and even conduits for addressing peoples' concerns about working conditions and pay. Recall Daniel's claim in chapter 8 that discussions with Filipino-American troops—who told him and other Filipinos that they were "getting screwed" by Serka—were the catalyst for the successful series of strikes against the company in 2004. Another example cited by former PPI employees at Balad is Brigadier General Oscar Hilman, who was in charge of base security from April 2004 to March 2005. Domingo, for instance, told me that Hilman would regularly come to PPI's mancamp, asking to eat adobo with workers, and speaking with them in Tagalog. He also played a central role in convincing Filipinos at Balad to continue working after a 2004 mortar attack on PPI's mancamp killed Raymond Natividad and wounded four other Filipino laborers.

A second factor behind Filipinos' relatively high status on bases, at least in the first year of the occupation of Iraq, was the Philippines' initial membership in the troop contributing "coalition of the willing." Manny remembers that Filipinos had badges with more privileges, such as the ability to shop at PX stores, than Indian and Bangladeshi coworkers due to the Philippines' coalition status. This is echoed by Angel and Domingo, who were part of the first batch of PPI employees to arrive in Iraq in October 2003. Angel, who worked in Baghdad, recalls, "We were not TCN . . . we were part of the coalition" and therefore "entitled to everything that the military was entitled to: DFAC, MWR, PX." According to Angel, Filipino workers in Baghdad lost these privileges "right after Arroyo left [the coalition]." When I asked if he knew why this happened he replied, "Yes,"—they were told that it was Arroyo's fault. Domingo remembers that the word "coalition" was written on his first badge, consequently he and other Filipinos on the base were given "full access, because we are allowed to go to DFAC, MWR." General Hilman even sustained these privileges after Arroyo's withdrawal of troops in 2004, making him a "hero of Filipino contract workers" at Balad.[13] According to Domingo, these privileges were only rescinded when Hilman left in spring 2005 and his successor forced PPI to rebadge all of its Filipino workers. Carlos, who jumped to a job with the private security firm Special Operations Consulting-Security Management Group (SOC-SMG) almost immediately after arriving in Iraq in early 2004, told me: "When I went to work for SOC they told me you can get your own CAC card because you are coalition. Then I got it and all the privileges . . . see, it says valid 2004 to 2006. After that we were not allowed to get a CAC card. With this there was much privilege. They treated you like a soldier when you wore that. Like an American." When the CAC card expired in 2006 he had to be rebadged and in the process experienced a loss of privileges, which in his humorous recounting was the first time he felt like a TCN with similar status to workers from South Asian counties. "I heard the term TCN when I renewed . . . Not before that. Then I asked, 'We are third country national? We belong like Indians, Pakistanis, Sri Lankans? Oh shit, we belong with those guys!?' [laughs] Damn!"

Hidden behind Carlos's joking concern that he "belonged" with "those guys" are nationally essentialist stereotypes circulating among Filipinos that discursively construct them as better workers than their South Asian counterparts. Moreover, as was frequently claimed in interviews, this superiority is recognized by both U.S. personnel and military contractors. Several factors make Filipinos ideal workers, I was told, the most important being the ability to speak English. Christian, for instance, claimed that soldiers preferred working with Filipinos in

Serka due to the language barrier with other nationalities, including their Turkish supervisors.

> CHRISTIAN: Filipinos are much different from Turkish [workers]. They can't speak English and understand. Only, "Yes/no. Yes/no."
>
> ME: So your Turkish supervisors would have to turn to Filipinos to translate?
>
> CHRISTIAN: Yeah, yeah! He would need help to translate from us. That's why Filipinos on U.S. bases are a priority. They [the military] want Filipinos.
>
> ME: And you were aware of this?
>
> CHRISTIAN: Yeah. They know that when speaking we can understand them. Not [like] other nationals, like Indians that [*he pantomimes an Indian yes/no head shake*].

In addition to language, several people cited Filipinos' supposed natural industriousness and flexibility. Gina, who spent nearly a decade in Afghanistan working in administrative positions for several different contractors, told me, "If you talk to some Americans . . . they like a Filipino, because [a] Filipino is hardworking, [a] Filipino, when you give instructions, only one time, they get it, they do [it]. What you want them to do, they will do it perfectly." Manny also discussed the superiority of Filipino workers in U.S. soldiers' eyes.

> MANNY: They [U.S. soldiers] would tell us about other job openings. And usually [it] would be [an] increase in pay. But it was up to you if you accepted or not. And if you don't accept no problem. They always offered the first opportunity to Filipinos. If no Filipinos then Bangladesh, Indians.
>
> ME: Why in your view did they offer to Filipinos first?
>
> MANNY: Filipinos are good workers. They take their jobs seriously. You do it your best. But other countries . . .

Most remarked upon was Filipinos' supposed cleanliness compared to South Asians. One recruiting agency owner, Gloria, focused on this quality when explaining to me why her company preferred Filipinos as workers. "They can communicate. Then, no smell. Very clean, take a bath . . . When they're in the dining facility, Americans want it clean. The KBR guys, they will check the dining facility. Our workers there, they say they will do like this [*wipes top of her desk with finger*] on the table and if it's dirty, they will really get mad. How can these Nepalese, Indian, Fiji guys do that?" Different standards of cleanliness also extended to conversations about life in company-run mancamps. This is especially the case for those who work in Afghanistan, where housing often consists of large Alaska tents filled with people from around the world, rather than segregated container units as was more common in Iraq. Consider the following exchange with Isko, who worked in both countries.

> ME: How many people [were] sleeping in a tent?
> ISKO: Fifty persons. All [from] around the world.
> ME: Did that cause problems?
> ISKO: It depends on how sloppy your roommates are.
> ME: Who were the sloppiest?
> ISKO: Indians. If you are tidy we will be fine. Kenyans are tidy. Kenyans
> are nice, and very industrious.

Following Anna Guevarra, I think it is useful to situate these comments—especially concerning Filipinos' supposedly inherent industry and cleanliness—within broader culturally essentialist and racialized discourses that "promote the Philippines as a *natural* source of *ideal* labor."[14] Such discourses are pushed by the Philippine state as part of its strategy of marketing labor for export. But their apparent resonance, among workers, troops, and contractors, also reflects more than a century of entanglement between Filipino labor migrants and U.S. military projects around the world.[15]

In contrast to the attention given to racial—and to a lesser extent, national—relations within the military's contractor workforce, when one reads news stories or academic analyses about those who support U.S. overseas wars it is hard not to notice the striking absence of female laborers. Indeed, one could be excused for thinking that no women work for military contractors in warzones as, with the notable exceptions of Sarah Stillman's 2011 long-form article "The Invisible Army" in the *New Yorker* and Lee Wang's 2006 documentary film *Someone Else's War*, women's experiences—especially those from other countries—are almost nonexistent. Yet a considerable number of women have also worked on military bases in the Middle East and Afghanistan. The military's contracting censuses do not provide information on the gender breakdown of its workforce in CENTCOM so it is not possible to calculate their presence with any precision. But my research suggests that it is more significant than has been acknowledged to date. For instance, nearly 20 percent of the workers I interviewed were women—and this with no attempt at oversampling along gender lines on my part. Moreover, when queried about the presence of female workers on bases, Bosnian and Filipino interviewees provided estimates ranging from 10 percent to 25 percent of the TCN workforce. My sense is that the lower bound is probably more accurate as interviews and news accounts suggest that it is much less common for women from South Asian countries like India, Nepal, Sri Lanka, and Pakistan to work for military contractors than those from labor-exporting countries in other regions of the world like Bosnia, Macedonia, Kosovo, the Philippines, Fiji, and Kenya.[16]

Irrespective of the actual numbers there is an evident disconnect between foreign female workers' not insubstantial participation in the military labor market and their near total erasure in reporting on the subject. This is perhaps not surprising as there is ample research demonstrating that women's perspectives and voices are consistently marginalized in news reporting, whether traditional print journalism, social media, or online news sites.[17] My interviews suggest that this disconnect is also fueled by relative differences in the type of work that women and men perform on bases, with the former more likely to be found doing administrative tasks (such as payroll, property management, and human resources), working in laundry or billeting, or occupying service positions in MWRs, PX stores, and other shops. Most of these jobs—with the exception of service positions—are less visible to journalists than male-dominated work like construction and DFAC operations—the latter perhaps the iconic symbol of TCN labor on military bases.

So how does the absence of women from accounts of military contracting matter? One way is through the framing of research agendas, especially when it comes to gender and the military. There is a rich body of feminist scholarship, for example, that examines topics such as how contractors perform masculinity, the intersection of masculinity and race in discourses about private military security contractors, the masculinization of military markets and the state, and the role that contracting plays in reinforcing "male dominance in the military and security sphere."[18] As this list of topics suggests, most scholars who focus on questions of gender, contracting, and the military do so through the lens of masculinity. To a certain degree the predominance of masculinity as a conceptual frame reflects the fact that the vast majority of this research deals with private security contracting, which is more obviously gendered than support work.[19]

Another way in which female military workers' absence matters concerns the lack of attention paid to intimate relations on bases. This lack of attention is notable because over the past two decades scholars have increasingly turned their attention to the intimate ties that have shaped U.S and European imperial projects, from sex to domestic work to child rearing. Focusing on relations between colonizer and colonized, this research has examined the ways in which "intimate domains . . . figure in the making of racial categories and in the management of imperial rule."[20] The context of intimate encounters on military bases is different, both in its relative narrowness (primarily sexual relations) and isolation from occupied populations. Nonetheless, these encounters are also revealing in their own ways when it comes to relations between and among American service members, contractors, and TCNs.

Nearly every person I talked with, for instance, indicated that relationships between American troops and foreign contractors are extremely rare. This boundary is policed by military brass, prime contractors, and subcontractors, with punishment for those working with the latter being dismissal. In contrast, two women who spent time on European bases in Iraq and Afghanistan—such as Camp Bastion, which was located adjacent to Marine Corps-run Camp Leatherneck in Helmand Province—recalled that it was common for coworkers to date European soldiers. When it comes to relationships among contractors the rules appear to be more varied, depending on the base one works at or the company one works for. Gina, who worked with Supreme and Arkel, among other firms, recalls that there were no rules against dating at these companies "as long as the work wasn't affected." In contrast, at Victory Base Complex KBR and PPI were stricter about policing relationships, especially between the companies' employees. Mary, who worked four years at the base with PPI told me that "they [would] terminate you" if they caught you dating someone from KBR, and that the same punishment also applied to KBR employees. Consequently most people she knew dated other Filipinos working for PPI.

KBR's rules against relationships with subcontractor employees appear to have been put in place in part to discourage exploitative solicitations for sex by American contractors. If so, their effectiveness was limited according to Filipina workers in Iraq, who recall that such relationships were not uncommon, especially among those that worked in billeting. The following story told to me by Iris, a single mother who worked for PPI at Balad, is instructive. According to Iris, she and several other women at the base had a profitable "extra business" cleaning rooms outside of regular work time. "If some KBR [worker] wanted you to clean their room they would pay us $20. Once a week, cleaning." When word got around people began offering them money for sex.

> IRIS: You know how many American guys approached me and said, "Be my girlfriend and I will give you money each month?" And I said, "Sir, I came here to work, not to sell myself. You offer me this big amount of money, but I don't need to." And they said, "Why don't you accept this offer rather than cleaning rooms?"
>
> ME: Americans would just approach you like that?
>
> IRIS: [nods]. Say you are my manager. So one day you come to me, "Iris do you need something extra?" This is their approach. Some ladies they want to flirt so they use it, "OK, sir, I want this, can you buy it and I will pay you later?". . . . I am talking from my own experience.

My boss came to me one day and said, "Iris, do you need anything from the PX?" "No sir." So then next day he approached me, "Iris, do you have something that you need to send to the Philippines?" "No sir." Third time, he said "Iris, why always when I approach you, you are telling me you don't need [anything], you don't like [anything]?" "Sir, I respect you as my boss. But respect me as your admin. I don't intend to work with you just to get involved with you. If I like you, I love you, I will give myself for free. But no negotiation." This is the only way that you can take care of yourself, by not letting other people use you.

Iris recalls that some of her friends whose "wish was to find money" did have "boyfriends." Likewise, Flora, who was employed in billeting for PPI at Victory Base Complex, told me, "Sometimes other women had three or four [boyfriends] . . . doing it for money. They would get into a relationship with a person that would support them financially."

The most common reason people develop relationships on military bases, I was told, is to satisfy a need for companionship and connection. Joshua, who was with PPI in Iraq, poignantly explained that "loneliness" was the primary motivation because individuals "just want to have somebody to be loved." A number of people from Bosnia recall friends or colleagues dating and even marrying men from the U.S. While such relationships were often dismissed as transactional, Diana insisted that in her experience this was rarely the case: "It's just love. It's just destiny. Everybody is searching for love, for happiness." Iris's experience illustrates Diana's argument. Eventually she began dating an American working for KBR, the two getting close enough that they began to discuss marriage. But he returned to the U.S. and after a year the long-distance relationship fell apart. "For three weeks," she tells me, "I was crying."

This need for human connection is heightened by the nature of life—isolated, regimented, and dangerous—on military bases in warzones. Several people alluded to prison when trying to describe their experience on bases. Representative is Goran, who told me, "It's a work camp. It's like a big prison camp. No one's going to hit you and shit, but your life is programmed. You eat at this time. You go and see that, that, and that guy at the same time every day. And it's shitty." Daniel, who worked for Serka in Iraq, stated, "We were like prisoners . . . just eat, sleep, and work." Likewise, Adnan, who was employed by KBR in Iraq and by Fluor in Afghanistan, called bases a "voluntary prison" where "you are like a machine. Wake up. Work. Eat. Sleep. That's it." In such a context many, whether married or not, desperately sought out companionship on bases. As Mary remarked, "Once you get there, it doesn't matter if you're married or not. You're both single."

Alen, who worked for KBR and Fluor in Afghanistan, provided an example of this for me. "Something happened that started to shake our family at that time. I got involved with another girl [on the base]. I got madly in love . . . When I came back our marriage was a disaster, I wanted to go away, I wanted to leave and marry that girl." Eventually he reconsidered, realizing that this relationship was a product of his lonely, pressure-cooker life on the base. "I started to think about what I'm doing. Is this the right choice? Is that really, [the] real girl? What about my wife? What about kids? How would they do growing up without me? I left my son. I left my daughter. I left my wife. OK. I left what we created together. At that time, I recognized the truth. The truth was it wasn't the right choice. The truth was that all that I created in my head about that girl was just my creation. . . . I really was crazy at that time."

Several people I interviewed met their current partner while working on a base, or knew others who had done so. Sam and Anne met while working for Kulak in Balad. Despite company rules against relationships, they began dating. After several months she got pregnant—"That one, he's Iraq-made," she joked, pointing to their oldest child—and returned to the Philippines. Their situation was not unique, according to Sam. "A lot of people had a really good opportunity to find a good relationship—it doesn't really matter [whether] with a Filipino or a foreign national. Most of the people that we know ended up in a relationship." Tatijana's brother, Luka, who oversaw property management at several bases in Iraq, met his wife, Katrina, on a short visit to Fallujah. "I just met her [briefly]. I mean with some other friends. We drank coffee. That's it. . . . Then the questions, 'Where you work at?' This and that. We started to email each other and then plan the vacation together and then another one and that's it. I met her on a camp where I was just two days." Adrijan, who is from Macedonia, also met his wife, Danica, who is from Tuzla, in Iraq. They were both in unhappy marriages, he remembers: "I was already having problems back home, she was also having problems back home, so it's probably just . . . it just happened."

Intimate relationships on military bases are not without consequences. One is the strain it places on relationships back home, as Alen's, Adrijan's, and Danica's stories illustrate. Another is that some subcontractors instituted changes in hiring practices, limiting opportunities for women. For instance, in response to a number of pregnancies among its workforce, Serka instructed its recruiting agency in the Philippines to quit hiring women.[21] PPI, I was told, took a different tack and started to prioritize hiring older women, like Iris, Mary, and Flora, under the assumption that they would be less likely to get pregnant. According to Mary, "They hired old, old. They didn't want young women because they didn't want [pregnancies]." In addition to placing the blame for pregnancies on women,

Serka and PPI also refused to provide access to contraceptives. KBR, in contrast, provided condoms at its camp in Balad, according to Domingo.

Finally, a darker side to this story concerns sexual harassment and assault on bases. Few I talked with were willing to discuss this topic openly like Iris did, but it was alluded to several times. Here, for example, is how Tatijana responded to a question about sexual harassment and assault during her time at Victory Base Complex:

> TATIJANA: It's not easy when you think about it. You're in military base with all those soldiers around and sometimes you have to go back [to your housing] . . . eventually they installed this buddy system [so] that you couldn't walk by yourself. Initially women had to get escorts. If you were leaving, I don't know if it was after dark or after hours, or was it all the time. They changed it. It eventually became you can't walk by yourself pretty much at all. You had to think about that too.
>
> ME: Was this [sexual assault] fairly common?
>
> TATIJANA: I didn't have issues like that but yeah, there were cases and complaints. I guess it's all about being careful. Being aware of your surroundings. Nothing different than being around here.
>
> ME: Yeah. Except for you're on a base, so there should be some more sense of security, you would think.
>
> TATIJANA: Yeah, but when you think about it, the majority is guys, both contractors and military. Then you consider the heat, and people go crazy when it's hot. Yeah, it's a little bit maybe more intense when it comes to work [there].

As Tatijana points out, severe gender imbalances and a heavily masculinized working environment are two factors that contribute to the cases of sexual harassment and assault on bases.[22] Another is the battlefield environment itself, as military-funded research indicates that rates of sexual assault against female military personnel increase in warzones compared to stateside bases.[23] Due to a lack of comparable research it is difficult to tell just how pervasive a problem this is for female contractors (foreign or U.S. citizens), but Sarah Stillman's investigative reporting suggests this is a significantly underreported phenomenon that is also exacerbated by the military's unwillingness to police the behavior of its contractor workforce—whether the matter concerns trafficking, labor abuses, or sexual assault.[24]

HOME

> There is a social and economic impact on everybody. You know, it's like both sides of a coin. It's good but you pay [for it] in other ways.
>
> —Enis

The comforts of home and family loomed large in almost every interview I conducted with Bosnian and Filipino military laborers. This makes sense, as amidst divergent experiences working and living on bases, absence from home constitutes one of the few commonalities shared by TCNs. The communities they come from are also important—if overlooked—sites in which the effects of the wars in the Middle East and Afghanistan are felt, with the primary conduits being workers themselves. These effects are multiple, from the trauma of returning dead and wounded bodies to the injection of money that alters the lives and trajectories of households and towns, to the toll that this work has on personal relationships. In this book and elsewhere I argue that such space-spanning entanglements are reshaping the geography of war. Due to military labor contracting on a scale and scope unprecedented in U.S. history, numerous communities and states around the world seemingly unconnected to the country's wars are nonetheless profoundly impacted by them as the effects of violence radiate far beyond the immediate battlefields. I refer to this condition as the "everywhere of war."[1] Perhaps nowhere else is the everywhere of war so deeply felt and intimate as places where recruiting for this type of work is highly concentrated, such the Tuzla valley in Bosnia and the Pampanga region in the Philippines.

I orient this chapter on home around three themes. The first concerns specific effects of the U.S.'s overseas wars on communities and families in Bosnia and the Philippines, especially those with high concentrations of military laborers. The second focuses on workers' longing for family and friends while living a secluded life on bases halfway across the world, and how they communicate with those back home. The final section takes a different tack. In it I explore the question of how political, social, and economic contexts at home shape individuals' ability to adjust to life after military work, including retrospective perceptions of the upsides and downsides of such work.

Entanglements

Few events illustrate more directly the connection between military contracting and the everywhere of war than deaths of foreign workers and the reverberations they cause back home. This tragedy has struck the small Bosnian town of Lukavac twice. The first time occurred in June 2008 when Nedim Nuhanović, an electrical mechanic for KBR, was killed by a mortar attack on a small base along the Afghanistan-Pakistan border. Nuhanović had been in Afghanistan for just six months. Nearly two years later, in March 2010, Fluor employee Almir Biković, who had spent three years as a firefighter at Bagram Air Base outside Kabul, was killed in a rocket attack on the base. Prior to this he had worked for several years for U.S. peacekeeping forces in Bosnia. Both deaths dominated local and national news for several days and left distraught family and friends in their wake. Biković was an only child, while Nuhanović, who worked as a video technician and DJ prior to heading to Afghanistan, had planned to marry his long-term girlfriend while on R & R in July. On the day that Biković died the Bosnian portal bliskiistok.ba temporarily crashed as thousands of people flocked to the site to read the breaking news. Days later hundreds lined the cold, wet streets of Lukavac as his funeral procession passed by, just as much of the town had gathered to bury Nuhanović two years before.[2] Years after, their deaths still resonated in Lukavac and nearby Tuzla, with several people mentioning them in interviews. One day an individual I will call Ado, who was chatting with me about my research in Lukavac, informed me that he had also applied to work in Afghanistan for DynCorp, and in fact had been offered an Asian contract. In the end Ado, who worked as an interpreter for U.S. peacekeeping forces in Bosnia in the 1990s, declined. When I asked why, he replied, "Two guys from this city died, Almir and Nedim. One of them was engaged to a girl from my neighborhood. And then my sister and brother told me, 'Ado, this is not Bosnia, it is not Europe. Afghanistan is a different story.'"

Despite the risk exemplified by the fate of Nuhanović and Biković, thousands from Lukavac and Tuzla have worked in Iraq and Afghanistan over the past fifteen years, lured by the chance to earn some "bread." Such lucrative opportunities are few and far between in the local job market, especially for young people given the heavily industrialized region's postwar economic decline. Shortly before his death, for instance, Biković was able to buy his own apartment, which is rare for someone in their early thirties in Bosnia. He was not alone. Indeed, America's wars since 9/11 have had a noticeable impact on Lukavac's urban fabric. Driving into the town is like passing through a massive industrial gateway, as the road is flanked by Bosnia's largest cement plant on the right, and the sprawling Soda Lukavac soda ash production facility on the left. The town itself has a rundown feel to it, with the center dominated by drab, Yugoslav-era apartment complexes. The notable exceptions are several recently built, modern- looking apartment towers surrounded by parking lots at the southeastern edge of the town, which locals colloquially refer to as "Iraq" and "Afghanistan" due to the large number of units purchased by people who have worked for military contractors in those countries. Similar, newly built, apartment towers have also sprouted up around the outskirts of Tuzla.

Though not as visually arresting, neighborhoods and towns in the Philippines have been no less significantly transformed by military contracting. For instance, a handful of former PPI workers I talked with came from a rural *barangay* in the town of Lubao, in Pampanga. All had relatively new concrete homes with metal or tile roofs, which they were eager to show off. The following exchange with Angel is representative.

> ANGEL: It was a happy but scary time. I was happy because I was able to build this house and send my kids to school. Most Iraq workers built new houses. This one here [*points to house across the road*] is my brother's. He was working in the gasoline pumps, with the Turkish drivers.
>
> ME: So many [people] from Lubao worked over there. How has it changed the community?
>
> ANGEL: Before the houses were just small houses on stilts and wood. Now they are concrete. These are our peace of mind. And now there are water wells. A lot of children were able to go to school. People bought vehicles.
>
> ME: This looks like a prosperous village.
>
> ANGEL: That is because most went to Iraq. When we were in Saudi Arabia it was a small salary. It does not compare to Iraq. You cannot build a house like this if you are working in Saudi [Arabia]. You cannot send your children to private school. But we in Iraq sent our daughters and sons to private schooling.

Echoing Angel, Christian, who worked for Serka in Iraq, told me: "The earnings from Iraq were so huge. This was not our house, it was just a shanty before. Every time I sent money home so that when the rain comes we will have shelter. And when I come home I am so happy even though I have no money. All the money went right here to our house [a beautiful three-story house]. And some education assistance for my children. So when I go home I had nothing other than my last salary."

In several cases people insisted I take pictures of them in front of their new homes. Andrew, one of Angel's neighbors in Lubao, had me take the picture reproduced in figure 10.1. Wearing a Marine Corps T-shirt, he informed me that his house was *katas ng Iraq* ("fruit of Iraq"), a phrase I heard from others in his *barangay*.

As Angel's and Christian's comments indicate, another significant area that money from military work has been directed to—especially in the Philippines—is education. Specifically, this entails paying to send one's children to private schools, which are perceived as superior to poorly funded public education options in the county. Even more than housing, Filipinos I talked with stressed the importance of education opportunities afforded by their military labor. For Fred, who worked for four years with Serka in Iraq, education was his primary motivation for applying.

> ME: What was the discussion like with your family when you made the decision to go?
> FRED: I wanted to go because the twins were going to college. I knew there was a war there. But I wanted to sacrifice for the girls.
> ME: Had you worked abroad before?
> FRED: No, my first time. My family agreed with me. Because we needed money for college. I am only a high school graduate. That's why I want my kids to go to college.

Like Fred, Angel contrasted his education status—"I was only in high school"—with his three children who will be able to get "good jobs" due to their private college education. "My daughter is a nurse at a hospital. My second finished [her] foreign service degree. My youngest will graduate as a civil engineer." Angel spent six years without a break working for PPI in Iraq, prompting me to inquire if there was a point during this time that he wanted to go home. He replied, "Oh yeah. But if you go home early you cannot go back [because of the travel ban]. My daughter at the time was in college. And I was worried that she might not be able to graduate." Similarly, when I asked Flora if she is still happy with her decision to go to Iraq in 2004 she replied, "Yes," because "I was able to send my children to [private] school, even though I am a single parent."

The economic and social effects of military work on communities in Bosnia and the Philippines extend beyond workers' deaths and investments in housing and education. Michelle, the local recruiter for Serka who was responsible for

FIGURE 10.1. Andrew in front of his *katas ng Iraq* ("fruit of Iraq")

helping dozens of people from her *barangay* obtain jobs with the company, highlighted several more subtle effects this has had on her poor community. She
claimed, for instance, that "for the first time families were able to celebrate birth-
days for their kids [by going out for a meal at Jollibee's—a popular Filipino fast
food chain—or McDonald's] and invite their friends to the celebrations." More-

over, "With so many families building or expanding their houses with the money they were earning, many construction workers didn't have to go live in Manila or even farther away in the Philippines to find work. They could work where they lived, in the *barangay*." Finally, she told me, many former Serka workers have subsequently found good-paying jobs as chefs, bakers, or kitchen assistants with companies in the Philippines and beyond. When I asked why this was the case she said: "Because they have experience working for a U.S. company [showing me KBR certificates of appreciation and food safety given to her husband]. This is like their passport to the jobs, because the certificates for food safety are valuable, because the U.S. Army is very strict about food safety and preparation." Michelle's claims about the value of Filipinos' experience working for U.S. military contractors and their subcontractors stands in contrast to Bosnians' complaints about the devaluation of their work experience at home. But her point about remittances having effects that extend beyond immediate families does apply. Indeed, as I argued in chapter 4, for roughly a decade the influx of money from the war economy in Iraq and Afghanistan was able to counteract—to an extent—general economic decline in the Tuzla region by bolstering industries as diverse as real estate, construction, restaurants, auto sales, tourism, and retail.

Another entanglement is the impact that military work has on families, especially those with children. While money earned from this work can transform families' material and educational situation, those who stay behind have to bear the load of raising children and managing households on their own. Consider the following exchange with Michelle.

> ME: What was the hardest thing about this work?
> MICHELLE: He could not come home. He was not with us during vacation times, during Christmas, New Year's. For me, my children are growing up. And I am raising them as a solitary parent. That was hardest.
> ME: Did you ever ask him to come home?
> MICHELLE: Yes. When there was the explosion at the DFAC [in Mosul, in 2004]. Most of us here [in the *barangay*] told them to come home after that bombing.
> ME: How did that conversation go?
> MICHELLE: They first said, "Yeah, we might come home." And then later on they said, "No, we are staying."

Rosamie—whose husband worked for Supreme for four years in Afghanistan—told me that she barely had time to be lonely because "I was busy every day, going to the school, the market [and] carry[ing] on by myself with the kids."

In addition to increasing the burden of reproductive labor on those at home, being apart also causes strains on relationships. Zlatan, who got divorced shortly after returning home, told me this was a common occurrence among other former military contractors he knows, especially those who were gone for years. In his case, he recalls: "We didn't fight. We didn't argue. We were just sitting and talking just like you and me now. 'OK. This is not going anywhere. This is not it. We lost too much time.' I know lots of people that got divorced in this area here [Lukavac]. I don't know. I'm looking at it like why? Sometimes you win. Sometimes you lose. Pretty much, you can't have it both, it looks like. You can't be on the other side of the world and you have a family here. You're just losing time." Rena also blamed her time working in Iraq and Afghanistan for the collapse of her marriage: "It was a phantom distance. It just made us know that we can live without each other. In one moment he told me, 'We live good without you,' and that made me so pissed. I didn't live good. I didn't live good at all! 'You live good because I send you [money] and you live exactly how you want because I provide [for] you. You don't appreciate that'. . . . Those couple of words made me—well, of course to respect myself [she left him]. If nobody else will I am going to." To add insult to injury, she told me, "When I came back from Afghanistan—I was two years over there—I came back and found $1,200 in my bank. That's it. He wasted the money . . . like I was going to stay [in Afghanistan] forever." Echoing Rena, Manny summed up for me the consequences of working with Serka in Iraq in the following way: "I built my house when I was in Iraq. But my family was broken as well . . . Too much trouble. That is my experience. I lost money. I lost family."

"Your Life's Not Complete"

Losing connection with family back home was a concern for most people I interviewed. This was especially the case for those with children. Representative in this regard is Kenan, who worked for five years with Fluor in Afghanistan. When I asked him what he found to be the biggest challenge related to his work, he immediately replied, "Reconciliation with family . . . especially if you have small kids. I went to Afghanistan when my older boy was three. I came back when my younger boy is three, so basically nobody knows me." For Rena, being apart from her daughter was an ever-present sorrow that made it difficult to work and sleep.

> If I call her—we do shifts over there, first and second shift. When I do first shift, if I call her after the job, I could not sleep over night. I would stay awake all night long and crying. Then if I call her before [the] shift,

I will be looking bad when I'm working. I could call her on my days off to be able to cry all day long as much I want. We are really close. When she's sitting next to me, I always need to touch her. Touch her hands so I always play like this [*caresses one hand with the other*]. When I do this, it's so nice. She's sitting next to me and all, always touching each other and I always play with that part of the hand. In Afghanistan during the night, it happened that I dream I do that. That wake me up and that's it. It's no sleeping anymore. You can't sleep. You['re] just thinking about sad things.

In the end, Rena, told me, "It was actually her only that [was] pulling me back home," not the relationship with her husband, which had slowly dissolved under the strain of years of being apart.

In the Philippines several people I talked with had parents who had also migrated abroad for work when they were young. Rowel brought this up unprompted while discussing the increase in privileges—particularly R & R every six months and the use of personal cell phones and computers—that occurred when he jumped from PPI to Parsons.

> ME: What would you do during your R & R's?
>
> ROWEL: We just keep, stay home, and then like Sunday go church. After that one, take a rest a little bit and then go to the mall. Spend my time with my kids. Some relatives is coming because they know you came from abroad, they got start coming, visiting you, then drink, cooking, barbeque, always doing get together. Not like my father, because my father was abroad also. During his time you can contact your families only by writing [letters], and then . . . sending in the post office. But not like now [where] we got computer, we got cell phone. Our communication is easy.
>
> ME: When you grew up, you didn't see your father much because he was abroad?
>
> ROWEL: Yeah . . . Since I was like, probably like four years old, [when he] start working in Saudi [Arabia].
>
> ME: For most of your childhood? What was that like growing up without your father?
>
> ROWEL: You feel like it's not complete. Your life's not complete because your father is not here. By the time that you need your father, you need some advice. Not like other people walking on the street, you see they are complete. They are working together with their families, father and mother. Then you saw them. You're going to miss your dad. You feel incomplete in your life. It's too hard.

When I asked Rowel if he was worried that his children would also feel "incomplete" due to his long-term absence, he replied: "I think it's better than my father because at that time you cannot talk to your father on phone, on a computer. You only talk to your father when in person. Now it's easy to communicate with your family on a computer, on a cell phone because cell phone they got camera, computer also. When you talk to them it's like it's with you, you get together. You feel like they are with you." As I discuss in the next section, Rowel also justified his choice by asserting that it would lead to a better future for his children, one which would not require the same kind of sacrifices made by him and his father.

Many people I talked with suggested that one of their central concerns was hiding details of work and life on bases from family. Specifically, this involved minimizing information about attacks, casualties, or dangerous working conditions so their families would not worry about their safety. For instance, when I asked Sead what he would talk about when chatting with his parents over Skype he replied:

> SEAD: Most of the time about what's happening here, that is Bosnia. You know, you cannot tell your mom there was a rocket [attack], you know. And when they hear the siren [signaling an attack] . . .
>
> ME: When you're calling?
>
> SEAD: Yes and they say, "Hey, what is that?" And you say, "They have practice for something. We need to go." So you just say, "Bye, see you tomorrow." But sometimes they watch the TV and see in Afghanistan is killed twenty people and they call tomorrow and ask, "What was that?" and you say, "Oh, nothing happened, it's not here, it's far from here." But I remember [one time] . . . the Taliban guys shot our container and there was [a guy] on Skype with family. I remember that. I mean, I mean some pictures are never going to go from your head like that, and he died in that place. . . . So in that time I want to go home, so I go in the office and say, "Hey, please, I want to go home. What do I need to do?" But, you know, there was nothing that happened to you so you think, "Oh, maybe nothing will happen again." So when you go sleep in your tent, tomorrow morning you're a different man. Just put that behind you and go forward. So I stayed. And after that I stayed two years.

Later in the interview Sead explained that he hid the details of this attack from his parents until he returned home, because he knew that they would have begged him to come home if they found out.

The most extreme example of hiding information I was told came from Grace, the single mom who was working illegally in Dubai in 2005 when a PPI recruiter convinced her to go to Iraq.

ME: What did your family say when you told them?

GRACE: They didn't know. Actually, they don't know that I'm going in Iraq a long time. I didn't tell [anyone] until one of my cousins, it was two years [later], yeah, that we spoke in Messenger. He said, "Hey, I went to your place, your address [in Dubai] that you gave and you're no longer living there."

ME: Wait, you were working in Balad for two years without telling your family?

GRACE: Yes, they don't know I was in Balad. I was pretending [to be] in Dubai. I get a lot of pictures [of Dubai] to show them. "Oh, this is my picture from that time." I just made basically . . . I just basically edit [pictures] in a computer and said, "Oh, this is the day that I . . ."

Even after her family found out about her move to Iraq she deflected concerns, responding to questions about life on the bases by saying "It's OK, it's easy. All is free . . . [you] don't have to worry."

Deflecting concerns from family and friends about the dangers of military work in warzones is understandable. But it is not without consequences. In fact, the emotional distance that Zlatan and Rena spoke of in the previous section is fueled in part by such silences. As Srdjan put it to me, "There's a big gap of say six or seven years" of life separating him and his wife. In an attempt to bridge this gap she eventually bought a copy of *The Kite Runner*, by Khaled Hosseini, which helped him open up about his time in Afghanistan. "I was living in that neighborhood in Kabul. And then she reads the book and comes to me and then I tell her what I saw with my eyes! Stuff like that, simple things . . . we'll discuss it from time to time."[3] Srdjan's case appears to be the exception as most I talked with found it difficult to discuss with family their experiences on military bases, even after returning home.

Afterlives

What is left after the money is gone? I posed this question to a group of former PPI employees one afternoon in Lubao. "Kids who have [a] better education," replied Angel. "Yeah, and they get a good job," followed up Chris. Others pointed to the many new houses and improved infrastructure in the community. With few exceptions, in fact, people I talked with in the Philippines felt that military work elevated their families into a better situation than before they left. This perspective is noticeably divergent from the more equivocal assessments of Bosnians. On its face this constitutes a puzzle. While working for military contractors offers Filipinos relatively better pay than similar jobs with civilian firms in the Gulf region, the differential is not enormous. And this work is arguably more precarious and dangerous, especially following the imposition of travel bans. In contrast, Bosnians working for prime contractors have been able to earn wages that are not just substantially greater than Filipino military laborers, but extravagant compared to the few job opportunities available at home. They have also experienced more opportunities to gain promotions and raises. So how are we to understand this discrepancy? The answer, I argue, lies in the different political, social, and economic contexts Filipinos and Bosnians experience upon returning home, which shape their adjustment to life after military work and retrospective perceptions of the upsides and downsides of such work. In this final section I examine these differences by comparing the afterlives of military work in the Philippines and Bosnia.

One significant difference concerns social expectations and perspectives on transnational labor migration. As discussed in chapter 3, since the 1970s the Philippine state has promoted labor export as a development strategy. In the intervening decades millions of Filipinos have headed overseas for work. According to the POEA there were 2.4 million OFWs in 2015. But registered OFWs are just a fraction of the overall number of Filipino citizens living and working abroad, which the government estimates to be as many as 10 million people—or roughly 10 percent of the country's population.[4] What this means is that labor migration is a relatively common experience for Filipino families. Indeed, several people I spoke with indicated that going abroad to pursue military work represents a continuation of previous labor migration to the region for individuals (as was the case with Angel) or across generations (as was the case with Rowel). Consequently, challenges associated with labor migration—from the burden on those who stay behind to strains on familial relationships—tend not to be suffered in isolation as more often than not relatives and family friends are experiencing similar issues. Michelle, for instance, highlighted one time that she and other spouses in her *barangay* intervened when the wife of a Serka worker was being profligate with money sent home by her husband. To provide another example, Gina left to work in Afghanistan when her daughter was six months old. When I replied that this must have been difficult she disagreed, replying that her mother was happy to look after her daughter. And shortly before I interviewed her in 2015, her daughter, who is now a teenager, encouraged her to apply for military work again if she wanted, saying, "You want to work again, mama, overseas? It's OK for me because I can manage . . . my grandma and I can manage."

Transnational labor migration is not just a common choice for Filipinos looking to improve the lives of their families, it is also socially and politically valorized. This is perhaps best exemplified by the government's promotion of migrant workers as *bagong bayani* (modern-day heroes), a phrase first used by President Corazon Aquino in 1988. In the thirty years since Aquino's invocation of *bagong bayani*, the Philippine state has diligently labored to "manufacture heroes" out of migrant labor.[5] Beginning in 1989, for instance, the POEA began sponsoring an annual Bagong Bayani Award that "seeks to recognize and pay tribute to our OFWs for their significant efforts in fostering goodwill among peoples of the world, enhancing and promoting the image of the Filipino as a competent, responsible and dignified worker, and for greatly contributing to the socioeconomic development of their communities and our country as a whole."[6] Central to *bagong bayani* discourse is the notion of migrants' experience of hardship and suffering, which sanctifies them as heroes of their communities and the Philippine nation. As Anna Guevarra observes, this aspect of the *bagong bayani* discourse is rooted in "Catholic ideals of sacrifice, suffering and martyrdom." Since these are culturally familiar and important values, "when the state invokes them, Filipinos understand and respond accordingly."[7]

While working abroad is both common and celebrated in the Philippines, the social context of labor migration in Bosnia is rather different. To begin, Bosnians' choice to work as military migrants is not valorized by either society or their government. The state does not track labor migration, and provides little support to workers or their families when crises arise, leading to a sense of social isolation. According to Srdjan, this isolation is amplified by the effects of working in a warzone—especially after surviving the war in the early 1990s—as illustrated by the following exchange.

> SRDJAN: Believe me, it took a couple of months to wind down, settle. And figure out, there is someone sleeping next to me. My wife. First couple of months I kept continuously waking up at 5:20 in the morning. Where am I? OK, I'm home, nice. Just to get your organism back [to] civilian life, and how should I say it? Socializing. I got together with my boys in this bar, [called] Oscar. We grew up together, went through the war together, everything. So they were so glad I am back, and happy for me. But it was a month after working and one time, "Srdjan, why are you so quiet?" "Guys you just talk your talk, I need time to take in everything." You know what I am saying? It was just like I was in my world trying to figure out shit. And it took some time, believe me. People changed.
>
> ME: This seems to be little difference [psychologically] with soldiers.
>
> SRDJAN: I would say it is like a 85 percent match. Because practically you were wrung through the same shit. Except for shooting. You were not in direct combat, but for everything else you were like a U.S. soldier.

You were in the same convoy, on the same chopper. In the same shit day in, day out.

ME: Is there anything in Tuzla, Lukavac, [other] local communities, support networks that have been developed?

SRDJAN: No. None that I know of. But I remember we were joking, just for those PTSD [post-traumatic stress disorder] guys, or those who miss it, we should build ourselves a camp outside town somewhere.

ME: So you can pretend to be locked up [on a military base] again?

SRDJAN: Exactly! [*laughs*] Just to have a feeling how it is. And simulate the same situation!

ME: That's some typical Bosnian black humor!

SRDJAN: So another price for that [work] is being without your family, totally separated, with strangers who come from different place. War going on. And I still don't know what damage has been done to my brain or my soul, but I am trying to keep my mind straight. And I think I am pretty good with it so far [*knocks on wood*]. But some people can't.

Srdjan was not the only person I talked with who suggested that the cumulative effect of living through the war in Bosnia and then working in a warzone exacted a psychological toll. Fedja worked as a labor foreman for KBR at Tallil Air Base for only four months before resigning. When I asked why, he replied: "For a lot of reasons. It was the third war in my really short period of life. I had the whole war here, had shit-tons of bad situations in Bosnia. And then I worked for almost six years in something like a SWAT team [a special police force]. And then I went to Iraq and there was a lot of shelling and stuff . . . The day before I went home there was eight guys in my camp [Tallil] killed. We had incoming shells and one of them hit a jeep and killed three MPs instantly on the spot, and five guys from India." Fedja then told me that when he arrived back home on his first vacation and saw his family he said to himself, "The money is not worth it . . . It's [working in Iraq is] too much for me," and decided that he would not return to Iraq.

A second factor concerns the history of labor migration in Bosnia and the broader "Yugosphere." While there is a tradition of temporarily migrating abroad in search of better pay and opportunities dating back to the Yugoslav period, the most common pathway for Bosnians has involved traveling to Western Europe, whether as a formal *gastarbeiter* (guest worker) or finding work in the informal economy, usually through personal connections with earlier migrants who have permanently settled in another country. Compared to these options, working for military contractors is a rare and relatively incommensurable form of labor migration. Moreover, while the skills and experience that people accumulate working in Europe often lead to new opportunities upon return, those who have been employed by military contractors report that the opposite is true in their case.

The primary reason that Bosnians are equivocal about military work, however, has to do with the general condition of precarity in Bosnia, from high unemployment and economic insecurity to corruption and divisive ethno-nationalist politics, and struggles to regain a sense normalcy in the aftermath of war and displacement. As we saw in chapter 4, for many this is encapsulated in the expression, "facing the reality of this life here." This "reality" profoundly colors people's perspective on the future, fueling pessimism that money, experience, or skills gained through military work will translate into a better life going forward, and placing emphasis on what has been sacrificed in a futile attempt to better one's life. Here is how Enis articulated this pessimism:

> When U.S. troops pulled out of Bosnia, Bosnians went with U.S. troops [*laughs*]. So still they were supporting their families back home—buying apartments, resolving existential needs, buying cars, getting a guy to paint my house, whatever. You help the local economy. And now that's going out too. And now what? You got a bunch of people that got back home and are now scratching their balls and what the fuck are they going to do? Do I invest in the local economy which is ruined, and with a questionable outcome of my investments? Do I try to go back again to some warzone? And for how long can you take it? Especially if you got kids. I know guys who haven't seen their daughters—just Skype and R & R—and then you lose them, there's a gap right there. It's like, "Yeah, my daddy is on a TV, and that's it." There is a social and economic impact on everybody. You know, it's like both sides of a coin. Its good, but you pay [for it] in other ways. You pay for it by being separated from your loved ones, or PTSD. There is a huge impact on the local population here. And meanwhile, unfortunately, things got worse in Bosnia, or our hometown [Tuzla].

When asked to assess her decision to work in Afghanistan with DynCorp, Rena offered an even more blunt and negative assessment: "You know how I describe my two years in Afghanistan? I wasted two years of my life because I didn't make it while I'm going over there. OK, I get some experience. What am I going to do with it? Nothing. I went to provide [a] better life for my family and I didn't." The primary long-term consequence of this work, she concluded, was the tension it placed on the relationship with her daughter's father, which eventually led to their separation.

Relations with family and friends, others told me, become even more strained as money drains away after people come back home and struggle to find work. Ivan explained this to me in the following way: "Money gets spent. Money, every day it's less and less and then you start fight[ing] with your wife on money a lot. Those are the downsides. Lot of marriages getting divorced. When you have a little more money, you start feeling beautiful. People like you all the time if you have money. Then after that, you're going to feel the real life. Over the night, people are going to start turning their head away like they don't know you anymore. Like you don't have money, they don't need you." Likewise, Sead argued that adjusting to straightened financial circumstances is the biggest challenge most military workers face when returning to Bosnia: "You know in our country they say *najgorije nemate pa imate* (it is the worst to not have after you have had). You know, because, you don't have money and you live with that. But when you live and you don't have money, you get some money, and then lose that again—don't have money—it's very bad. It's killing you in your head." Due to the depressed economy and difficulty in finding work—even work that pays Bosnian wages—nearly everyone who returns home, he claimed, wrestles with this decline in status.

Not all in Bosnia are so pessimistic. Kenan told me that he spent the first three months back "just watching TV" but then "one day you wake up in the morning and say to yourself, 'Yeah, well this is a different reality, let me swim in this reality now.'" He then decided to invest in a construction company—"two excavators and two trucks"—attributing his optimism to experience working with Fluor.

> Well, I have more confidence in myself. You have to understand I wasn't in Bosnia for five years, so I kind of forgot how the system works here and how much people are suffering because [of the] economic situation . . . I was in my dream world like Alice in Wonderland. I'm coming from Afghanistan to spend twenty-one days here [every R & R], so my only aim is to have fun with my family and I have money to support that, so I don't give a heck about the political situation, I don't give a heck about economic [situation]. Those people, they don't exist for me because I'm stuck in my world. Now, when you come back, you start to awake. You can see how the real life actually is, but because everything I went through to put myself in this life position and not in some other, I have a choice . . . . I don't see problems where most people see them. I'm above the lethargy, which is in every sphere of living in this country. It's in people's heads. It's on [the] street. It's everywhere. That's my benefit from [Afghanistan].

Like other long-term workers, Kenan is suggesting here that he picked up different habits and ways of thinking after interacting with American contractors and uniformed personnel for years. But unlike most—who emphasize the challenge this imposes on readjusting to life back home—he insisted that this has enabled him to stay "above," mentally, the precarious reality of life in Bosnia. That said, later in the interview he acknowledged that since he had only been home for six months when we talked, his optimism might fade over time.

Whereas Bosnians are generally pessimistic, Filipinos I spoke with tend to be optimistic that military work will lead to a better life for their families. This is reflected in the ubiquity of references to the future during interviews. Rowel, for instance, explained to me: "You can live [here], but it's not like—I mean everyday life you can survive over here but the future of your family you cannot reach over here especially if you don't have a business. Our choice is going out of the country, travel abroad. . . . The only thing I think is if I got to stay home the future of my family and my kids is, I cannot give them a good future. That's the feeling—that I'm going to be strong, stay outside [working on the base]. Just keep putting in my mind the future of my kids." When I asked Mary about the conversation she had with her daughters before leaving for Iraq she replied, "I told them that [it's] for their future. That's right. 'If I don't go there, how you can finish your study?' I told them, explained to them." One person I talked with, named Edwin, worked abroad for more than thirty years, twenty with construction companies in Saudi Arabia and eleven with military contractors in Iraq and Afghanistan. He admitted that he felt homesick many times over the years, "But if you think about your family it will pass" because "this is their future. I am doing this for them."

Rowel's comment about being able to "survive" in the Philippines, but a better future being out of reach for his family if he did not decide to work abroad, reflects two widely held assumptions among Filipino military laborers I interviewed. First, most people who have pursued this line of work come from relatively poor and underprivileged families and communities. Like Rowel, they view working abroad as the most realistic chance to escape life at the economic and social margins. This squares with Amanda Chisholm's research on Nepalese security contractors who also see military work as a short-term sacrifice that will provide a better future for their families.[8] Second, education—specifically their children's education—is seen as the primary mechanism that will allow families to move from the margins. The "good future" that Rowel believes he has secured for his children is based on their ability to go to a private school. As noted above, private schools are perceived as superior to public ones in the Philippines, both in terms of the education they provide and the opportunity for social advancement that they afford. One useful way to think of this is a process of converting economic capital to cultural capital, as suggested by the following analysis of the link between remittances and private education in the Philippines:

When economic capital is circulated back to the Philippines, it becomes convertible to other forms of capital. A common use of remitted funds is the education of siblings, children or other relatives. In this way, economic capital is converted into cultural capital, which forms an investment in the sense that such cultural capital will, in the future, itself yield economic capital. The ability to keep children in school and, in particular, the ability to send them to prestigious schools or colleges, also constitutes an important conversion of economic capital into social capital as parents develop new networks among a higher status section of society, and children develop friendships, social ties, and alumni networks with a similarly elevated cohort.[9]

To return to the question that I began this section with, then, for Filipino military laborers the money earned working on bases in Iraq and Afghanistan is not gone. Instead it has—ideally—been transformed into other forms of capital that will benefit their families for generations to come. This optimism is based upon an understanding of life in the Philippines as socially stratified but also relatively fluid if one can acquire the educational and cultural capital necessary to achieve a middle-class life. No such optimism exists for Bosnians, who have little hope for a better future due to pessimism about the suffocating "reality of this life" in their country.

EMPIRE'S LABOR

Empire involves more than pushpins on a map. It is made up of human activities—a network of situated practices that . . . sculpt geographies in their own image.

—Josh Begley

On October 4, 2017, four SOF personnel were killed in an ambush near Tongo Tongo, a remote village in western Niger. That the U.S. was carrying out military operations there—and subsequent revelations that roughly 800 personnel were located in the country at the time—came as a shock to most Americans, including members of Congress. In an interview with the NBC news show *Meet the Press* days after the attack, South Carolina senator Lindsey Graham, one of the more knowledgeable members of Congress concerning foreign policy, admitted, "We don't know exactly where we're at in the world, militarily, and what we're doing."[1]

For those who follow military contracting trends on the continent the large U.S. presence in Niger was less surprising. As noted in chapter 5, in early 2013 the Air Force established a drone base in the capital, Niamey. Three years later, according to contracting documents, the base had "a steady state of 200 to 250 personnel a day."[2] In 2014 the Pentagon moved its airlift contract for casualty evacuation, personnel recovery, and search and rescue support from Burkina Faso to Niamey, indicating a significant shift of SOF forces to Niger.[3] That same year the military announced that it planned to establish a second drone facility in Agadez, a desert city more than 700 kilometers northeast of Niamey. Satellite imagery indicates that the still-under-construction base will have a footprint that is larger than Camp Lemonnier in Djibouti by area. Finally, in 2015 (or possibly earlier) the U.S. established a secret SOF base next to the massive uranium mines in Arlit, near the Algerian border.[4] As the designated contractor for AFRICOM under the LOGCAP IV contract, Fluor has provided logistical support for each of these bases. In fact, one can roughly track the inexorable increase in the U.S.

military's presence in the country by monitoring the steady flow of positions advertised at the company's LOGCAP job opportunities website.[5] Less than two weeks after the deadly ambush, for instance, Fluor advertised several new positions at the SOF base in Arlit, including a plumber, a vector control specialist, and a food service supervisor.

This conclusion addresses the following question: How has the revolution in military logistics and contracting impacted the "American way of war"? Shortly after the invasion of Iraq in 2003, military historian Max Boot wrote an influential article in the journal *Foreign Affairs* arguing that technological advances were ushering in a "new American way of war." Whereas before the U.S. relied on numerical superiority in weapons and men to wear down opponents, Iraq—and the war in Afghanistan—demonstrated a new paradigm of warfare, one in which "quick victory with minimal casualties" and minimal cost is achieved through "speed, maneuver, mobility, and surprise."[6] Fifteen years on, with trillions of dollars spent, thousands of U.S. personnel killed, tens of thousands more wounded, and hundreds of thousands of civilians dead, this prediction reads like a cruel joke. Instead of quick and painless victory, the "war on terror" grinds on, with little change in policy other than an expanding roster of countries in which the U.S. now carries out operations.

Indeed, if there is any defining characteristic to the American way of war in the present day it is the unboundedness of its spatial and temporal registers.[7] Spatially, this "everywhere war" is nearly unlimited in its ambition, extending even to space and cyberspace. One of the more striking aspects of this spatial unboundedness is the ubiquity of "war in countries we are not at war with."[8] The growing U.S. military presence in Niger, and deadly violence that has accompanied it, is an excellent example of this element of the everywhere war. The temporal counterpart to everywhere war has received even more attention over the past two decades, with America's continuous military operations since 9/11 variously characterized as "endless war," "infinite war," the "long war," and the "forever war."[9] Again, the dramatic increase in U.S. military presence in Niger in recent years, and Africa more generally, suggests that there is no end in sight to America's spatially and temporally unbounded wars. This too was acknowledged by Senator Graham in his interview with *Meet the Press* when he stated, "This is an endless war without boundaries and no limitation on time or geography."

If U.S. military ambition—and hence its imperial foreign policy—is now defined in large part by this peculiar combination of everywhere and forever war, what enables this state of affairs? Like Boot, most observers stress technological innovations. Technology is important, and undoubtedly part of the story. But technological wizardry alone is an insufficient basis for prosecuting boundless war. As I have argued in this book, the ability of the U.S. to project force, continuously and on a planet-wide scale, depends as well upon the immense logistical resources it can bring to bear. This includes both logistics spaces, including its global network of bases, and logistics labor, which is now drawn from around the world. Indeed, it is scarcely an exaggeration to argue that logistics "holds empire together across time and space."[10]

Moreover, technological changes—such as the ongoing "robotic revolution"— and increased reliance on foreign labor (and foreign military surrogates) over the past decade and a half, represent two sides of the same coin, which Martin Shaw has identified as "risk transfer war."[11] According to Shaw, this "new Western way of war" is centrally concerned with "managing relationships between political risks (to politicians) and life-risks (to combatants and civilians)" by transferring them onto foreign societies and bodies.[12] Above all this entails minimizing casualties to Western soldiers. The utility of drones and other robotic systems, such as Explosive Ordnance Disposal machines, in facilitating the transfer of risk by minimizing casualties on the battlefield is recognized.[13] Less so is the concomitant risk transfer role played by contracting, though as noted in the introduction contractors constitute roughly one-third of the casualties suffered by U.S. forces and its associated civilian workforce in CENTCOM since 9/11. In both cases the transfer of risk and casualties onto foreign bodies serves to dampen domestic opposition to the pursuit of boundless war. Here, Cynthia Enloe's observation that "the wheels of militarization" are "greased . . . by popular inattention" is instructive, as few things disrupt inattention to the U.S. military's boundless wars more than the deaths of American soldiers.[14] Put another way, the new American way of war is a product of changes in both technology and military contracting.[15]

It is necessary, then, to push back against accounts that argue that technological innovations are heralding a new form of warfare in which machines reduce the need for military bodies and labor.[16] Emblematic of this view is Ian Shaw's "predator empire" thesis. According to Shaw the spread of drone operations signals that "American empire is transforming from a labor-intensive to a machine- or capital-intensive system." Consequently, "the new face of the U.S. military's empire has far fewer human faces."[17] On the surface drones appear emblematic of innovations toward small-footprint, technologically sophisticated and machine-intensive military operations that enable the U.S. to extend its reach across the globe. However, as my discussion in chapter 4 concerning the extensive logistics sites and labors that supported a tiny drone outpost in Ethiopia from 2011 to 2015 suggests, it is a mistake to succumb to this machinic seduction. Instead, a more accurate observation is that "distributed and labor intensive" drone operations "do not so much do away with the human but rather obscure the ways in which human labor and social relations are configured."[18]

There are further problems with the argument that technological advances are lessening the importance of military labor. First, the drawdown of troops in Iraq and Afghanistan that Shaw highlights reflected, in large part, a strategic shift by the Obama administration away from war in the pursuit of regime change, occupation, and counterinsurgency to a focus on counterterrorism. This shift was clearly articulated in the 2015 *National Security Strategy* report which states: "We shifted away from a model of fighting costly, large-scale ground wars in Iraq and Afghanistan in which the United States—particularly our military—bore an enormous burden. Instead, we are now pursuing a more sustainable approach that prioritizes targeted counterterrorism operations, collective action with responsible partners. . . . Working with the Congress, we will train and equip local partners and provide operational support to gain ground against terrorist groups."[19] Counterterrorism lends itself much better to smaller military footprints, especially when combined—as noted in the report—with a liberal reliance on military labor contributed by local allies and proxies, such as the thousands of Chadian, Malian, Cameroonian, and Nigerien forces that are providing the bulk of troops for counterterrorism campaigns in the Sahel region of Africa. This point is echoed by Brigadier General Donald Bolduc, the former commander of SOCAFRICA, who observed in 2016 that effective counterterrorism operations on the continent are not possible "without enablers, robust logistics, intelligence and airlift, *host nation forces* and international partners" (italics mine).[20] The primary mission for many U.S. SOF operators in Africa, in fact, is training host country military forces. These foreign "human faces" should not be discounted when accounting for the military labor of U.S. empire. Moreover, strategic priorities change. If the U.S. initiates another war in the name of regime change—as has been advocated by some foreign policy hawks with regard to Iran or North Korea—it will once again be accompanied by large-scale military deployments.[21]

Second, the U.S. military remains highly dependent on labor, but this dependence is obscured by reliance on foreign workers, whose presence, as this book argues, is typically overlooked. Shaw, for instance, cites the reduction of U.S. troops in Afghanistan to a "skeletal force" of nearly 11,000 by the end of 2014 as representative of the reduction in labor accompanying counterterrorism operations.[22] These troops, however, were accompanied by more than 39,000 military contractors in the country at that time. The vast majority of these were TCN and Afghani laborers providing logistics support.[23] Moreover, the nearly 4:1 ratio of contractors to troops in Afghanistan at the end of 2014 was substantially greater than any previous period in U.S. history. Nor was this a temporary anomaly. A year later more than 30,000 contractors were still supporting a U.S. force just short of 9,000 uniformed personnel.[24] Furthermore, these numbers do not represent a full accounting of the labor involved in continuing military operations in Afghanistan. Missing from the data are thousands of truck drivers, stevedores, and warehouse employees in Pakistan and various Central Asian countries that move supplies to bases in Afghanistan, contracted airlift transporting workers and troops in and out of the country, and back office staff of military contractors and subcontractors working in office parks in Dubai. Indeed, what is most striking with regard to military operations since 9/11 is not a reduction in labor that sustains them, but its changing composition, from uniformed and American to civilian and foreign. In short, military labor still animates U.S. empire, but where it comes from, and how it is obtained, has changed significantly over the past two decades.

The parallel here with earlier European empires' dependence on military labor performed by colonial subjects to sustain their imperial projects is evident. Consider the following observation: "A durable imperial system can afford to make only moderate military demands on the 'home' population. The British empire would never have been so popular for so long with the British public if every single soldier who policed that empire had to be recruited in Britain. Thus the Indian army helped to make the empire politically palatable in Britain by reducing the demand for British soldiers and taxes."[25] This point holds true today. Contracting reduces the demands of America's pursuit of boundless war with regard to deployed personnel and casualties, thus reducing political risk. But whereas European empires primarily relied upon the labor of colonized peoples, the sources of the U.S. military's present-day workforce are more diverse. In addition to enrolling former colonial subjects like Filipinos, workers are drawn from sites of previous interventions, including the peacebuilding missions in Bosnia and Kosovo, and transnational capitalist labor mobility circuits, such as the massive labor import regime established by Gulf petro-states.

Tracing these heterogeneous military labor pathways, the histories that have produced them, and the various political, economic, and social entanglements that radiate back out along them, reveals critical—but less-known—contours of the U.S. military empire. It also bears witness to the fact that this empire is inextricably linked with the lives of the global army of labor whose thankless toil it depends on.

Acknowledgments

This book could not have been written without the generosity and assistance of a number of people and organizations. I want to first thank everyone in Bosnia and the Philippines who welcomed me into their homes, communities, and businesses, and patiently took the time to discuss their experiences with me. I regret that I cannot mention by name (due to confidentiality reasons) a number of individuals in Bosnia to whom I am especially indebted due their time and guidance that helped shape this project during its early stages. Deserving of special recognition is Lucille Quiambao, my guide to the world of military laborers in the Philippines. In addition to being a logistics magician, Lucille has an uncanny knack for putting people at ease, even when discussing difficult topics. And as a Kapampangan with extensive experience exploring this phenomenon, her suggestions and insights were always spot on.

Though rarely acknowledged, research and writing is always shaped by encounters with people and places with no direct connection to a project. Several deserve mention here. First, the amazing people, music, and gin and tonics at the aptly named—and sadly departed—Caffe Galerija Bunt ("Uprising") in Tuzla offered a most welcome place to unwind after a day of fieldwork. Second, most of this book was written at a number of cafés and bars in Westwood, including Elysee Bakery and Café, Rocco's Tavern, and the Hammer Museum Café. Third, special thanks is due to the University of California, Los Angeles (UCLA) Faculty Center, in particular all the underpaid and underappreciated staff that make it a wonderful spot to eat, drink, think, and work on campus—especially Rodolfo, Miguel, and Celestino.

David Phinney and David Isenberg provided valuable advice at an early point in this project. So too did William "Bill" Stuebner. Thanks to those who read and provided comments on all or parts of the book at various stages, including Jennifer Mittelstadt, Mark Erbel, John Agnew, Gail Kligman, Jackson Lears, Robert Mobley, Michael Mann, and Leiba Faier. I also appreciated comments from the manuscript's reviewers at Cornell. I would like to extend my gratitude to Jennifer Mittelstadt and Melissa Feinberg for inviting me to a Center for Cultural Analysis symposium at Rutgers University that provoked stimulating discussions about military contracting. Much thanks is also due to Joseph Blatt, Nerve Macaspac, James Walker, and Vernon Wessel for assistance with the research, and to

Matt Zebrowski, the UCLA Geography Department's cartographer, and my collaborator on the book's maps and figures.

This is my second book with Cornell University Press, and with Roger Haydon as editor. From previous experience I knew Cornell would be the best home for this project. True to form, Roger—and all the other staff at Cornell—were a pleasure to work with again.

Portions of the book draw upon previously published material in the journals *Geopolitics* and *Territory, Politics, Governance*. Thanks to these journals, and to my coauthor of the *Geopolitics* article, James Walker, for permission to use this material. Research for this project was facilitated by a generous fellowship from the Hellman Foundation for which I am grateful. Additionally, this book is freely available in an open access edition thanks to the Toward an Open Monograph Ecosystem (TOME) initiative and the generous support of Arcadia, a charitable fund of Lisbet Rausing and Peter Baldwin, and of the UCLA library. I very much appreciate Sharon Farb's enthusiastic help in facilitating this.

Finally, this project has benefited from love, support, food, and conversations shared with family and friends, especially my parents, Lori, Danica (no longer so *mala*, but still daddy's *princeza*!), Bill and Katy, and, last but not least, Camille—whose friendship means more than words can adequately express.

This book is dedicated to all those who toil in, as Lee Wang has evocatively put it, someone else's war.

Notes

1. MILITARY CONTRACTING, FOREIGN WORKERS, AND WAR

Epigraph: Julie Greene, "Builders of Empire: Rewriting the Labor and Working-Class History of Anglo-American Global Power," 3.

1. Embassy Kathmandu 2004.

2. Cam Simpson (2018) has chronicled—in beautiful and evocative detail—the lives and deaths of the Nepalese men, as well as repercussions on families back home and the subsequent decade of lawsuits in U.S. courts. For more contemporaneous reporting on these events, see Dhruba and Rohde 2004; Bell 2004; T. Miller 2006.

3. Schwartz 2010. A copy of the November 2008 report, as well as all of the other quarterly censuses, can be downloaded at http://www.acq.osd.mil/log/ps/centcom_reports .html. For a map of CENTCOM's AOR, see http://www.centcom.mil/images/stories/unified -command_world-map.jpg.

4. Roberts 2014. For a detailed analysis of one USAID contractor project in Afghanistan, see Attewell 2017.

5. Raw contracting census data from Iraq (see chapter 2), for example, indicates that the number of contractors in that country increased from nearly 137,000 in 3rd quarter 2007 to more than 149,000 in 2nd quarter 2008, and reached its peak at the end of 2008. This corresponds with the peak in the average monthly number of U.S. military personnel in Iraq, which also occurred in 2008. For more on this, see Belasco 2009.

6. This graph is based on data from all published quarterly censuses beginning in 2008. Reports can be found at http://www.acq.osd.mil/log/ps/centcom_reports.html.

7. See, for example, Scahill 2008.

8. On sovereignty and the monopoly of violence, see Avant 2005; Verkuil 2007; Krahmann 2013; McFate 2015. On state control and accountability, see Singer 2008; Isenberg 2008; Bruneau 2011. On effectiveness, see Dunigan 2011. On ethical and moral implications, see Pattison 2014; Eckert 2016. One significant exception to this policy-centric focus is an emerging literature that examines the intersection of gender, race, and masculinity with private security contractors. See, for example, Joachim and Schneiker 2012; Higate 2012a; Eichler 2013, 2014, 2015; Stachowitsch 2014; Chisholm 2014a; Chisholm and Stachowitsch 2017.

9. Runstrom 2010, slides 29–30.

10. Fontaine and Nagl 2010, 9. Fontaine and Nagl derived these figures from an analysis conducted by the U.S. Army's Center for Military History on behalf of the Commission on Wartime Contracting for Iraq and Afghanistan (CWC), which is the most widely cited analysis of wartime contracting ratios.

11. This book focuses on logistics contracting by the U.S. military. Relatively less research has been devoted to logistics contracting by its Western allies, though they too increasingly rely upon contractors to support overseas operations, such as the NATO-led International Security Assistance Force mission in Afghanistan. This reliance on contractors is especially true of the United Kingdom, which has moved toward the U.S. model for logistics contracting in recent years, as evidenced by the introduction of its own multiyear Contractor Logistics contract patterned after the LOGCAP program in 2004. See Kinsey

and Erbel 2011. For more on UK and NATO logistics contracting, see Kinsey 2009; Cusumano 2018.

12. Singer 2009; Gregory 2010.

13. Ferris and Keithly 2001.

14. Quoted in Farrand 2006, 1.

15. For more on the operational link between RMA and logistics outsourcing, see Erbel and Kinsey 2018.

16. On urban battlespaces, see Graham 2009. On war in borderlands and ungoverned spaces, see Bachmann 2010; Mitchell 2010; I. Shaw 2013; I. Shaw and Akhter 2012. On lawfare and lawful targets, see Gregory 2006; Khalili 2012; Weizman 2012; Jones 2015, 2016.

17. Gregory 2011.

18. Moore 2017.

19. For data on U.S. military casualties, see http://icasualties.org. For information on contractor deaths, see Department of Labor, Office of Workers' Compensation Program n.d. The Department of Labor is required by the Defense Base Act to track overseas contractor injuries and deaths for the purposes of providing compensation. It should be noted that the number of contractor casualties—especially injuries—is likely higher than the data indicate as it is incumbent upon companies to report these figures and there is evidence of underreporting by firms, especially subcontractors.

20. Jaymalin 2009.

21. Embassy Kuwait 2004e.

22. Embassy Kuwait 2004c.

23. Embassy Kuwait 2004d.

24. Embassy Kuwait 2004e, 2004f.

25. Tyner 2005, 109.

26. Pangilinan 2004.

27. R. Kaplan 2003.

28. Vine 2015, 3. See also, Lutz 2008. This said, as Catherine Lutz (2006, 595) points out, it is important not to forget that the U.S. is a settler colonial state that also still maintains a number of overseas island colonies, such as Puerto Rico, the Northern Mariana Islands, Guam, and the U.S. Virgin Islands.

29. Go 2011, 9. For more on informal empire, see M. Mann 2012, 18–19.

30. Singer 2008, 97.

31. Bender and Lipman 2015, 1.

32. Greene 2015, 36.

33. Greene 2015, 1. For more on this emerging field of inquiry, see Greene 2016. For prominent examples of this research, see Greene 2009; Poblete 2014; Lipman 2008. Following Neil Smith (2003), I view the American Century as beginning with the events of 1898—including, most pertinently for this story, the beginning of U.S. colonial rule in the Philippines—though the phrase was first coined by Henry Luce more than four decades later.

34. Stoler 2016.

35. Bender and Lipman 2015, 5.

36. Filkins 2008.

2. FROM CAMP FOLLOWERS TO A GLOBAL ARMY OF LABOR

Epigraph: Robert Gates, *Commission on Wartime Contracting in Iraq and Afghanistan*, 20.

1. World Health Organization 2014.

2. UN 2014. For more on the rise of international securitized responses to infectious diseases, see Davies 2008.

3. Obama 2016.

4. DoD, Office of Inspector General 2015.

5. Conner 2015.

6. Sloop 2016.

7. Reibestein 2015.

8. Directorate of Logistics 2015.

9. Moore and Walker 2016.

10. Conner 2015.

11. Kinsey 2009, 11.

12. Mayer 1996.

13. N. Baker 1971.

14. Delo 1992, 178–79.

15. Wilson 2006, 51.

16. Wilson 2006, 135–37.

17. Reinhart 2004, 85.

18. Wilson 2006, 1.

19. Wilson 2006, 3, and chap. 3.

20. Waddell 2009, 99.

21. Huston 1966, 286.

22. Huston 1966, 287; Bankoff 2005; Jackson 2014, chap. 1.

23. Huston 1966, 675.

24. McGrath 2007.

25. Mauldin (1945) 1968, 78–79.

26. For an excellent discussion of the tensions between "grunts" and "REMFs" in Vietnam, see Lair 2001, chap. 1.

27. Huston 1996, 384. Chung (2019) provides a detailed account of these labor dynamics.

28. Huston 1996, 646. Thousands of Korean workers and several Korean companies subsequently provided logistics support for the U.S. military in the Vietnam War, a fascinating parallel (and precursor) to the case of Bosnians who began working for the military in the postwar peacekeeping mission in Bosnia before moving on to new jobs in Iraq and Afghanistan. For more on this, see Chung 2019.

29. Huston 1989, 377.

30. For more on this, see Huston 1989, chaps. 15 and 16.

31. Sheehan 1967, 16.

32. "Vietnam" 1965, quoted in Zamparelli 1999, 7.

33. Dunn 1972, 27.

34. Raymond International and Morrison-Knudsen (RMK) held the original contract construction contracts. When demands ramped up too fast for the two firms to keep up with they invited Brown & Root and J.A. Jones Construction to join them. See Carter 2008, 158. Carter claims that RMK-BRJ was the "sole contractor" for construction projects in Vietnam, but this is inaccurate. PAE was a key construction contractor for the Army, while the Air Force relied on Walter Kidde Construction to help build its Tuy Hoa Air Base. For more on this, see Dunn 1972, 27; Traas 2010, 109.

35. Quoted in Carter 2008, 184.

36. Carter 2008, 185.

37. For more on ties between Johnson and Brown & Root, see Caro 1990.

38. Quoted in Carter 2008, 239.

39. For more on this transformation by Rumsfeld, see Beasley 2019.

40. Fontaine and Nagl 2010, 8.

41. Tillson 1997, 11.

42. DoD 1992, 727.

43. Among other activities, SCCC operated the dining facilities at Camp Bucca, a detainee prison in southern Iraq. In 1990 it provided food for American soldiers at the main logistics base in Damman. See Fialka 1990.

44. Dickinson 2011, 29.

45. As Jennifer Mittelstadt (2015) has shown, the privatization revolution left the military relatively unscathed during Reagan's administration. In fact, during this time there was a marked increase in the provision of social welfare services by the military, which were viewed as essential for the recruitment and retention of service members following the introduction of an all-volunteer force in 1973.

46. Department of the Army 1985, 1.

47. C. Smith 2012, 32.

48. Stollenwerk 1998.

49. Stollenwerk, 33.

50. Rasor and Bauman 2007, 246–47.

51. GAO 1997, 2.

52. C. Smith 2012, 34.

53. C. Smith 2012, 7.

54. GAO 1997, 4 and 11.

55. C. Smith 2012, 36.

56. Cahlink 2002.

57. D. McKenna 2002, 17.

58. D. McKenna 2002, 15.

59. In 2006 the Navy split its CONCAP program into two components, a Global Contingency Construction Contract, which focuses on overseas construction projects, and a Global Contingency Services Contract, which is used to provide logistics and base support services.

60. Cahlink 2003.

61. Grace 1984.

62. By the end of the Clinton administration the federal government had shed nearly 400,000 jobs from its payroll, and was spending 44 percent more on contractors than in 1993. For more on military and intelligence privatization during the Clinton administration, see Shorrock 2008, chap. 3.

63. Defense Science Board Task Force 1996. See the August 27 memorandum by Chairman Philip Odeen on pp. 5–6.

64. Voelz 2009; Shorrock 2008; Crampton, Roberts, and Poorthuis 2014.

65. Mittelstadt 2015.

66. Roberts 2014, 1034.

67. Despite assumptions that outsourcing leads to substantial cost savings, evidence suggests that the cost benefits are mixed at best. See Stanger 2009, 94–98.

68. Eichler 2014, 605.

69. Dao 2002.

70. Zamparelli 1999, 9.

71. McFate 2015, 43–44.

72. Palmer 1999.

73. Cohen 1997.

74. Bianco and Forest 2003.

75. FDCH E-Media 2005.

76. Raugh 2010, 174.

77. Serafino 2001; Department of State 2001.

78. Waddell 2009, 172. According to one estimate, by the late 1970s 75 percent of the Army's combat service support capabilities belonged to reserve components. See Stollenwerk 1998, 12.

79. A. Tyson 2006; Musheno and Ross 2008.

80. See Schooner and Swan 2012, and Propublica's investigative series at https://www.propublica.org/series/disposable-army.

81. Avant and Sigelman 2010, 245.

82. Hyndman 2007.

83. Defense Science Board Task Force 2014, 12.

84. When the 2004 visa ban crisis crippled transportation operations into Iraq, PWC was greatly affected as nearly 60 percent of its 1,500 drivers were from India and the Philippines. See Embassy Kuwait 2004c.

85. Chatterjee 2009, 133. According David Vine (2015, 218–20), who conducted an extensive analysis of Pentagon contracts performed outside of the U.S. between 2001 and 2013, the five largest contractors as determined by value of contracts were all logistics contractors: 1) KBR (LOGCAP), 2) Supreme Group (DLA), 3) PWC/Agility (DLA), 4) DynCorp (LOGCAP), and 5) Fluor (LOGCAP). In total he estimates these five companies earned nearly $80 billion.

86. Gordan 2014.

87. For a detailed overview of these claims, see C. Smith 2012, chap. 6.

88. For more on fixed and cost reimbursement contracts see C. Smith 2012, chap. 3, and http://farsite.hill.af.mil/reghtml/regs/far2afmcfars/fardfars/far/16.htm#P180_28325.

89. R. Brown 2009.

90. C. Smith 2012, 71.

91. C. Smith 2012, 45–46.

92. Grasso 2008, 17.

93. See, for example, Cha 2004; Phinney 2005; Simpson 2005a, 2005b; Rohde 2004.

94. Merle 2006.

95. Copies of these data, which were obtained following a FOIA request by journalists, are accessible at CENTCOM's FOIA reading room: https://www2.centcom.mil/sites/foia/rr/default.aspx. I have also posted copies of these quarterly censuses on my academia.edu webpage (https://ucla.academia.edu/AdamMoore).

96. Cleveland 2008.

97. Unfortunately it is not possible to compare these data on contracting personnel with the size of contracts. While there are government websites and databases (e.g., fbo.gov and usaspending.gov) providing information on military contracting and expenditures, they are incomplete. Moreover, in the case of subcontracting, the legal principle of "privity of contract" means that agreements between prime contractors like KBR and their subcontractors are subject to drastically lower degrees of transparency and oversight. Therefore, these data provide, as far as I am aware, the most detailed information on subcontracting in contingency operations that exists to date. For more on this, see Tyler 2012.

98. Several accounts, including an earlier work of mine (Moore 2017), inaccurately identify GCC as a Saudi company. But legal documents from a lawsuit between GCC and KBR identify the former as a Kuwaiti firm. See Duroni 2013. Eventually GCC became a subsidiary of PWC. See C. Smith 2012, 89, and Project on Government Oversight n.d.

99. C. Smith 2012, 83–85.

100. Following the LOGCAP IV award, Fluor and DynCorp took over KBR's operations in Afghanistan, but for continuity purposes the military decided that KBR should continue to provide logistical services in Iraq under the LOGCAP III contract until the withdrawal of troops in 2011.

101. Peters, Schwartz, and Kapp 2015, 3.

102. The six geographic combatant commands are also known as unified combatant commands (UCCs). In total there are ten UCCs, six organized according to specified AORs—the geographic combatant commands—and four along a "functional" basis: Transportation Command (TRANSCOM), Strategic Command, Special Operations Command, and Cyber Command.

3. COLONIAL LEGACIES AND LABOR EXPORT

Epigraph: Alfred McCoy, *Policing America's Empire*, 18.

1. Chatterjee 2009, 7; Stillman 2011; Phinney 2005.

2. Kavinnamannil and McCahon 2011.

3. Quoted in Go 2007.

4. Though rarely mentioned, the full title of Kipling's poem is "The White Man's Burden: The United States and the Philippine Islands."

5. This quote comes from Justice Harlan's dissent in *Downes v. Bidwell*, one of the Insular Cases through which the Supreme Court decided the political and legal status of the newly acquired colonies and their peoples. The dissent can be found at http://www .supremelaw.org/decs/downes/Justice.Harlan.dissent.htm.

6. Kennedy is quoted in Eakin 2002. See also Go (2005), who argues that the Louisiana Purchase provided key political and legal precedents for overseas territorial possessions.

7. Poblete 2014, 48.

8. Poblete 2014, 106. For more on the recruiting process and experiences of Filipino laborers in Hawaii, see also chaps. 2 and 4.

9. Baldoz 2011, 57.

10. Sharma 1984, 584.

11. Bender and Lipman 2015, 11.

12. Baldoz 2011, 61–69.

13. A. Kaplan 2003, 3.

14. Poblete 2014, 2–4.

15. Fujita-Rony 2015, 212.

16. R. McKenna 2016.

17. Bureau of Insular Affairs 1905, 375 and 379.

18. In 1907 this comparison was suggested by an American officer following a visit to observe British forces in Agra, India. See Purviance 1907.

19. Lasker (1931) 1969, 61.

20. R. Miller 2004.

21. Quinsaat 1976, 108.

22. Ingram 1970.

23. Quoted in, Dodd 1968, 41. Remarkably, the original draft circulated by the U.S. military in 1946 also tried to claim exclusive jurisdiction over U.S. personnel for any offenses committed while off base. This was a step too far for both Philippine negotiators and DoS officials, who convinced the military to remove this language. For a detailed account of base negotiations, amendments to the original agreement, and jurisdictional questions raised by a number of legal cases, see Berry 1980.

24. Simbulan 1983.

25. N. Williams 1987. According to a 1977 GAO report, pay for Filipino employees at bases ranged from 54 percent higher than prevailing wages for clerks to 111 percent for security guards. See GAO 1977, 6.

26. Zulueta 2012.

27. On Guam, see Flores 2015; Woods 2016. Wake Island, according to one account, "hummed with activity" between the 1950s and 1970s, including a "large contingent of Filipino employees" brought in by a U.S. contractor. See Gilbert 2012, 310.

28. Mynes 2010.

29. Bandjunis 2001, 194. For a detailed history of the U.S. military's presence at Diego Garcia, see Vine 2009. Presently, roughly 2,500 contractors work at Diego Garcia, the vast majority of them Filipinos paid as little as $2,200 a year. See McQue 2017.

30. Mynes 2010.

31. On U.S. empire and its overseas bases, see especially Chalmers 2004; Oldenziel 2011; Vine 2015.

32. Gregory 2006, 411. On Guantanamo, see also Kaplan 2005. On Diego Garcia, see also Vine 2009.

33. Legarda Jr. 1955.

34. Rodriguez 2010.

35. Government of the Philippines 1974, arts. 17, 19, 20.

36. Battistella 1999, 230; Tyner 2005, 37.

37. Government of the Philippines 1974, art. 22.

38. Philippine Statistics Authority 2016a.

39. De Vera 2017.

40. Paddock 2006. For a full analysis of Arroyo's rhetorical positioning of OFWs as the Philippines' primary global export, see Serquina Jr. 2016.

41. Duaqui 2013.

42. Orbeta Jr. and Abrigo 2009, 3–4.

43. Philippine Statistics Authority 2016b, fig. 4.

44. Birks, Seccombe, and Sinclair 1988, 267–68.

45. "Gulf Countries" 2016.

46. Dacanay 2004.

47. Quote comes from transcript of 2005 interview conducted by Lee Wang and Lucille Quiambao.

48. Woods 2016, 133.

49. R. Rodriguez 2010, xvii.

50. R. Rodriguez 2010, 1.

51. Choy 2003, 65.

52. Excerpts from this speech are cited in Choy 2003, 115–16.

53. Woods 2016, 132.

54. Woods 2016, 134–135.

55. The embassy's remarks are cited by Woods 2016, 143. See also Flores 2015.

56. Woods 2016, 149.

57. Glassman and Young-Jin 2014, 1176.

58. Central Intelligence Agency (CIA) 1969b, 3–4.

59. Tregaskis 1975, 235.

60. CIA 1969a, 136–37.

61. "Agreement" 1968.

62. United States Court of Appeals, Fifth Circuit 1992.

63. For a full list of bilateral labor agreements, see POEA n.d.

64. Tyner 2005, 3.

65. Celoza 1997, 100–103.

66. Arnold 2003.

67. Cited in Tyner 2005, 88.

68. Kammerer 2003.

69. O'Connell 2003.
70. "U.S. Military Logistics" 2002.
71. Mynes 2010.
72. Conde 2004.
73. PPI's award language is cited in Chatterjee 2009, 146. AES proudly notes their award on their website (www.angloeuropean.com.ph/#).

4. THE WAGES OF PEACE AND WAR

1. In 2006, when Goran quit his police job, 600 Bosnian marks was roughly the equivalent of $375 per month.
2. Raugh 2010, 163.
3. Disagreement over who would be accorded control over the Brčko area nearly scuttled the peace negotiations in 1995. At the last minute a deal was made to have the issue resolved through international arbitration within a year of signing the Dayton Peace Agreement. A final decision on Brčko's status was ultimately delayed until 1999 when the arbitral tribunal declared that the entirety of the former *opština* (a Yugoslav unit of local government similar to a municipality or county) would become an autonomous District. For more on this, see Moore 2013.
4. Gallay and Horne 1996, 9, cited in Dowling and Fleck 1999, 9.
5. "News Briefs" 1996, 3.
6. A December 1996 paper on privatization from the Center for Naval Analyses claims that Brown & Root alone hired 6,700 workers in its first year of operations, though it does not indicate the source of this figure. See Stafford and Jondrow 1996, 5. Brown & Root was the largest PMC in Bosnia, but it was just one of many U.S. and Bosnian firms supporting the peacekeeping mission, which makes 10,000 a conservative estimate in my view. A similar hiring boom—and subsequent migration of workers to the Middle East and Afghanistan—occurred in towns in Kosovo and Macedonia near the massive Camp Bondsteel base established by the U.S. military as part of the Kosovo Force peacekeeping mission. For more on the Macedonian context, see K. Brown 2010.
7. Jennings 2010, 231.
8. C. Baker 2014.
9. Jennings 2010; Jennings and Boas 2015.
10. The most notorious examples of the emergent sex industry in northeast Bosnia were the brothels and trafficking operations at the Arizona market near Brčko. For more on sex trafficking and peacekeeping in the Balkans, see Mendelson 2005. For more on the Arizona market, including the relationship between it and U.S. peacekeeping forces, see Moore Forthcoming.
11. See C. Baker 2014, 94.
12. Soriano 1996, 12.
13. Tokach 1997, 9.
14. Jasarevic 2014, 262.
15. Pargan 2009.
16. DynCorp's classification is evident in multiple company documents I have acquired.
17. Fluor's tier system was explained to me by several former workers. Data on the pay differentials comes from a September 2009 company document titled "FGG Contingency Operations: Salary Structure—Tier II/Tier III/Tier IV." Copy on file with author.
18. Pargan 2010.
19. Another sign of Bosnians' lower status while working with DynCorp is that the company allowed workers to take leaves just twice a year, and only paid for the cost of the flights for the first leave—unlike Fluor and KBR, which paid for the flights for all three granted leaves.

20. C. Baker 2012.

21. On the existential dimensions of precarity, see Ettlinger 2007.

22. On the importance of investigating the various political and institutional contexts involved in the production of precarity, see Waite 2009.

23. Jansen 2006.

24. Knežević 2017.

25. As Asim Mujkić (2016) notes, leaders of the 2014 protests also drew lessons from the JMBG (short for "Unique Master Citizen Number") protests the previous year, which centered on criticism of politicians' handling of a dispute concerning whether the country's identification numbers issued at birth should designate the ethnicity of citizens. The political deadlock lasted for months, resulting in thousands of citizens unable to obtain birth certificates, passports, and health insurance documents. For an excellent collection of analyses of the 2014 protests, see Arsenijević 2014.

26. See, for example, Kurtović 2015. On politics in Tuzla, see Armakolas 2011.

5. SUPPLYING WAR

Epigraph: Martin van Creveld, *Supplying War*, 1.

1. Thorpe (1917) 1997, 3.

2. ISR Task Force, Requirements and Analysis Division 2013.

3. Elish 2017. See also, A. Williams 2011.

4. Deptula 2010.

5. Scahill 2015.

6. Moore 2018, 337.

7. This information comes from online contractor résumés. Following Trevor Paglen (2009), I refer to this as "résumé intelligence" or RESUMINT. Online contractor and military personnel résumés offer especially rich veins of information on the various operations and activities conducted by the U.S. military around the world over the past decade. This said, I have decided not to provide links to individual résumés in the footnotes. There are two reasons for this. First, they can be easily altered. Second, the information they reveal constitutes breaches of operational security on the part of contractors and military personnel, therefore creating the risk of personal repercussions. However, full webpage PDFs of all pertinent résumés have been created and copies remain on file with the author. For more on RESUMINT, see Paglen 2009, 70–74.

8. U.S. Air Forces in Europe and Air Forces Africa A4A7K 2015.

9. Crampton, Roberts, and Poorthuis 2014.

10. This definition draws from Deborah Cowen's (2013, 8–9) discussion of logistics space.

11. Belanger and Arroyo 2012. On US military infrastructural investments in the Arabian Peninsula in the 20th century, see Khalili 2018.

12. Cowen 2013.

13. U.S. Transportation Command 2012, 2.

14. Quoted in Belanger and Arroyo 2016, 92.

15. U.S. Transportation Command 2016, 5.

16. For more geographical analyses of the inherent conflicts and tensions involved in logistics and the circulation of goods around the world, see Chua et al. 2018.

17. Semple 2005.

18. Santora 2009. As Lair Meredith (2001, chap. 1) demonstrates, luxuriously appointed wartime bases first appeared in the Vietnam War, with contractors also playing a key role in their construction and operations.

19. In six cases I have agglomerated data from bases or sites that were separated out in the raw data tables, but were adjoined and/or overlapping on the ground, and thus

functioned more or less as a single base. These are (1) Diamondback and Marez (surrounding Mosul Airfield); (2) Victory Base Complex surrounding Baghdad International Airport, including camps Victory, Liberty, Radwaniyah Palace, Mayberry, Cropper, and Slayer; (3) Kirkurk and Warrior (surrounding Kirkuk Air Base); (4) IZ sites, including the IZ complex, Tigris, and Freedom Rest; (5) Basra and Harper (surrounding Basra Airfield); and (6) a military training ground east of Baghdad that cycled through a variety of names during the occupation, including Shakoosh, Butler Range, Besmaya Range, and Hammer.

20. Englehart 2009.

21. Gisick 2010.

22. 301st Area Support Group (General Support Unit) Garrison Command 2005, 40–42.

23. Raz 2007.

24. Embassy Kuwait 2007d, 2009.

25. Embassy Kuwait 2009. As extensive as U.S. military use of Kuwaiti facilities was in 2009, it appears to have paled in comparison to the beginning of the war according to a 2003 cable that details a much more extensive presence, including the estimate that Kuwait had "set aside approximately 70 percent of its total land area for U.S. military training and bed-down" that year. See Embassy Kuwait 2003.

26. Embassy Kuwait 2005. Prior to the MOU a series of ad hoc agreements had governed border-crossing procedures. Due to a handful of disagreements leading to temporary border closures in the previous two years, negotiating the MOU was a priority for the military and DoS.

27. Embassy Kuwait 2007b. This cable, which was written before operations began at Khabari, stated that contractor conveys would still be subject to inspections. But this is contradicted by a 2009 military logistics article which states that "the Khabari Crossing, unlike Navistar, would operate as a throughput for convoys, not a staging yard. Staging would take place at other bases before heading for Khabari Crossing. At the new crossing, the previous convoy receptions, inspections, and consent procedures would no longer be used. Instead, civilian transporters would be issued a coalition crossing card—a plastic photo identification card with a bar code containing information linked to the Kuwaiti immigration and customs databases." See Walker 2009.

28. Embassy Kuwait 2007c.

29. Embassy Kuwait 2007a.

30. Embassy Kuwait 2009, 2006.

31. De Simone and Gauthier 2003.

32. Jet fuel (JP8) is also used by the military as fuel for M1 Abrams tanks, and for cooking, heating, etc.

33. McNulty 2009, slide 15.

34. Belanger and Arroyo 2012.

35. Belanger and Arroyo 2012, 55.

36. Embassy Islamabad 2009.

37. Pillsbury 2010.

38. McNabb 2009, 53–54.

39. Masood 2009.

40. Tellis 2011.

41. For a detailed analysis of U.S.-Pakistan relations, see Kronstadt 2011.

42. The most comprehensive and updated data on drone strikes in Pakistan are produced by the Bureau of Investigative Journalism. See https://docs.google.com/spreadsheets/d/1NAfjFonM-Tn7fziqiv33HlGt09wgLZDSCP-BQaux51w/edit#gid=1000652376. For a detailed history of drone strikes in Pakistan from the early years to the peak of operations

in 2010, see B. Williams 2010. On the exceptional status of FATA in Pakistan, see I. Shaw and Akhter 2012.

43. Secretary of State 2007.

44. Following the closure of Manas by Kyrgyzstan in 2014—in part due to pressure from Russia—the military began using a Romanian air base, referred to as MK, near the port of Constanta as its primary transit center for troops entering and exiting Afghanistan. See Nickel 2014.

45. Kuchins, Sanderson, and Gordon 2009, 7.

46. Perlez and Cooper 2010.

47. For earlier instantiations of NDN routes, see Kuchins et al. 2009, 9–10; Cooley 2012, 44. On Russia's decision to close off the northern line, see Daly 2015.

48. Kendrick, Hawkins, and Swan 2012; Andzans 2013.

49. Whitlock 2011. Data on shipment costs come from a TRANSCOM document, dated June 21, 2011, that responded to questions submitted by Whitlock in advance of the above story. This document is part of a large batch of NDN-related material made public following FOIA requests that can be accessed at the command's FOIA reading room. See http://www.ustranscom.mil/foia/reading_room_arc.cfm#hideD.

50. Boyd 2015.

51. At the peak of operations in 2011–12, roughly 70 percent of NDN cargo entered Afghanistan through Uzbekistan. See Kuchins and Sharan 2015, 105.

52. Embassy Tashkent 2009.

53. Cooley 2008, 225.

54. Cooley 2012, 39.

55. Cooley 2012, 40.

56. Trilling 2011.

57. On this distribution process, see Rackuaskas 2008, 14 and 17. Multiple Bosnian contractors I interviewed mentioned the cooling yards in Afghanistan. A description of this process is also provided by Task Force Currahee 2014, 41.

58. McNulty 2009, slides 15 and 17.

59. Contract language cited in Tierney 2010, 10.

60. Rackauskas 2008, 17.

61. Tierney 2010.

62. Roston 2009. In addition to Tierney's (2010) congressional report, Roston's reporting was subsequently substantiated by an internal U.S. Army investigation. See DeYoung 2011.

63. Khan and Abbot 2012.

64. Sopko 2014, iii.

65. Anderson 2011.

66. Pietrucha 2012.

67. On the decision to avoid supplying military escorts for Afghan truckers, see McDonnell and Novack 2004.

68. Gregory 2012.

69. For a detailed account of the lily pad strategy in Africa and other parts of the world, see Vine 2015, chap. 16.

70. Greitens 2011, 267.

71. See, for example, King, Moss, and Pittman 2014.

72. In addition to U.S. military drone bases on the continent, the CIA also operates at least one drone facility, in Dirkou, Niger. See Penney et al 2018.

73. Moore 2016a.

74. Corrick 2012, 46–47.

75. Moore and Walker 2016, 697. Camp Gilbert was reportedly used as a staging site for SOF missions in Somalia. See Schmitt and Mazzetti 2008.

76. On the expanded size of Camp Simba, see https://www.neco.navy.mil/upload /N62470/N6247015R4007RFP_15_R_4007_Djibouti.pdf.

77. See Moore and Walker 2016, 698.

78. From 2007 to 2012 Creeksand flights flew from Burkina Faso and Mauritania, and also provided ISR coverage over Mali and Niger. The Tuskersand operation (beginning in 2009) was part of a multinational campaign against the Lord's Resistance Army. Tuskersand was based in Uganda, with flights providing ISR coverage over parts of South Sudan, Central African Republic, and the Democratic Republic of the Congo. See Moore and Walker 2016.

79. Langley 2010.

80. AFRICOM 2010.

81. D. Rodriguez 2015.

82. For more on LOGCAP contracting and the use of LOGCAP contractors, see Moore 2017.

83. A copy of the solicitation is on file with the author.

84. On training at Camp Singo, see Whitlock 2012b.

85. Cornella et al. 2005a, ii.

86. Cornella et al. 2005b, viii. The number of CSLs in Africa mushroomed from a total of four in 2005 (Senegal, Ghana, Gabon, and Uganda) to thirteen in 2011 (Senegal, Gabon, Uganda, Ghana, Algeria, Botswana, Kenya, Mali, Namibia, Sao Tome and Principe, Sierra Leone, Tunisia, and Zambia). See Cornella et al 2005b and Ploch 2011. An unclassified 2018 AFRICOM briefing obtained by Nick Turse (2018) suggests that the number is now similar (twelve) but their composition has evolved. However, in my view the AFRICOM briefing functions more as a disinformation device than a true accounting of U.S. military presence on the continent. The three largest U.S. drone bases on the continent (Niamey, Chabelly and Agedez), for example, are labeled as CSLs, while two others (Garoua and Bizerte) are absent from the map. Moreover, a key SOF base in Kenya (Manda Bay) is listed as CSL even though its runway is too short to accommodate most large military and civilian transport planes. Finally, in responding to Turse's inquiries the Pentagon refused to acknowledge whether or not its tally is exhaustive—which it clearly isn't. Consequently figure 5.4. highlights only sites that have been confirmed as CSLs through previous sources.

87. See Dickey 2013. According to a 2015 military presentation, at least two of the Operation New Normal Marine staging bases (Libreville and Accra) are supported by LOGCAP contractors. See U.S. Army 2015, 32, 33, and 65.

88. Seck 2015.

89. On Ghana, Senegal, Gabon, Niger, and Spain, see Seck 2015; on Uganda and Djibouti, see Reif 2014; on Italy, see Vandiver 2014.

90. On strategic airlift channels, see Moore and Walker 2016, 698.

91. Turse 2017.

92. D. Rodriguez 2016.

93. On Mali, see Whitlock 2012a; on Burkina Faso, see Campbell 2015.

94. Reeve and Pelter 2014, 27.

95. Vine 2015, 97.

6. ASSEMBLING A TRANSNATIONAL WORKFORCE

1. This definition of migration infrastructures comes from Xiang and Lindquist 2014, 122. For more on the concept of "migration infrastructures"—particularly in Asia—see Lin et al. 2017; Hirsh 2017; Lindquist 2017.

2. The prominent role played by recruiting agencies, or labor brokers, is not unique to the Philippines. Indeed labor brokers are a critical—perhaps *the* critical—node of migration infrastructure for labor-exporting and -importing states in Asia, a fact that has generated increased scholarly attention in recent years. For more on this, see Lindquist, Xiang, and Yeoh 2012; Molland 2012; Kern and Muller-Boker, 2015; Lindquist 2012, 2015, 2017.

3. Tyner 2003, 23, 85.

4. Guevarra 2010, 92. For a more expansive account of the recruiting process, see chap. 4.

5. Tyner 2003, 90.

6. Castaneda 2004.

7. It is worth noting that following the imposition of travel bans in 2004, Serka evolved toward a body shop, expanding its services for the military in Iraq from food service to ice plant operations, water purification treatment, laundry services, and administrative support.

8. Phinney 2005.

9. These quotes are from a 2011 deposition of KBR's former vice president of accounting and finance, Government Infrastructure Division, William Walter. See C. Miller 2012.

10. Quote comes from transcript of 2005 interview conducted by Lee Wang and Lucille Quiambao.

11. Castaneda 2004.

12. Quote comes from transcript of 2005 interview conducted by Lee Wang and Lucille Quiambao.

13. Embassy Manila 2004a, 2004b.

14. This said, a handful of people I interviewed—such as Rena and Srdjan—were hired by other military contractors in Iraq and Afghanistan during this period.

15. Michels 2008.

16. For the original articles, see "Regruteri" 2017; Brkić 2012; Slavnić 2013.

17. Claudio Minca and Chin-Ee Ong (2016) provide an interesting examination of an earlier example of the use of hotels in Amsterdam to facilitate the transnational movement of labor. The broader observation about the underexamined geo-economic and geopolitical significance of hotels comes from Lisa Smirl's (2015) work on the spaces of aid, and from Sara Fregonese and Adam Ramadan's (2015) call for more focus on the geopolitics of hotels.

18. Morrison 2016; Smirl 2016.

19. For an account of the robbery, see Agarib 2015. One of the most notorious gangs in the world, the Pink Panthers are estimated to have made off with $500 million in jewelry over the years. Their exploits were the subject of the 2013 documentary film *Smash & Grab* directed by Havana Marking. For more on the gang see, Simon 2014.

20. Bonacich and Wilson 2008, 3.

21. Cowen 2013, 1.

22. Cowen 2013, 2; italics in original.

23. Alderton et al 2004. For more out flags of convenience, labor outsourcing, and the decline of labor standards in the shipping industry, see Bloor and Sampson 2009.

7. DARK ROUTES

1. Embassy Kuwait 2004a.

2. Quotations in this and the subsequent paragraph come from Embassy New Delhi 2004.

3. See Embassy Kuwait 2004b.

4. See, for example, Human Rights Watch 2009; Amnesty International 2016.

5. Black and Kamat 2014.

6. McCahon 2011.

7. See Department of State 2011.

8. Isenberg and Schwellenbach 2011.

9. Mayberry 2007.

10. Owens 2007. For more on First Kuwaiti's trafficking of workers and other labor abuses committed on the Embassy project, see Phinney 2006.

11. Owens 2007.

12. Stillman 2011.

13. The name of the employee has been redacted to protect his identity. Copy of contract on file with the author.

14. American Civil Liberties Union (ACLU) and Allard K. Lowenstein International Human Rights Clinic, Yale Law School 2012.

15. Li 2015, 127.

16. Appel 2012, 697.

17. Simpson 2018, 260–61.

18. Cha 2004.

19. See Propublica's remarkable series of articles on the disposable army of contractors in Iraq and Afghanistan at https://www.propublica.org/series/disposable-army.

20. See Wise 2013. For more on subcontracting and labor abuses, LeBaron 2014. The transference of risk onto workers—especially from Asia, Africa, and Latin America—can also be seen in the realm of private security contracting, not just logistics. For more on exploitative labor conditions for security contractors from the Global South, see Gallaher 2012; Chisholm 2014a; Eichler 2014; Thomas 2017.

21. Department of the Army 2016, sect. 1, p. 3.

22. McCahon 2011.

23. Nolan 2010.

24. POEA 2011.

25. Harris 2006.

26. Ruiz 2012.

27. Jaymalin 2010.

28. Ellison 2014.

29. Mount 2008.

30. Embassy Baghdad 2009.

31. ACLU 2012, 47.

32. Whelan and Gnoss 1968; Thrasher 1993.

33. Nagl and DeMella 2011, 15.

34. See *Federal Acquisition Regulation (FAR) 52.215-2 Audit and Records—Negotiation* at https://www.law.cornell.edu/cfr/text/48/52.215-2.

35. See *Federal Acquisition Regulation (FAR) 52.222-50 Combating Trafficking in Persons* at https://www.law.cornell.edu/cfr/text/48/52.222-50.

36. Black and Kamat 2014.

37. See *Federal Acquisition Regulation (FAR) 52.222-50 Combating Trafficking in Persons* at https://www.law.cornell.edu/cfr/text/48/52.222-50.

38. Carp 2007, 48.

39. Carp 2007, 48.

40. CWC 2011b, 159.

41. Darwin 2008, 442.

42. Scully 2001, 5.

43. Raustiala 2009, 19.

44. Waits 2006.

45. Warren 2012.

46. Finer 2008, 264.

47. These quotes are excerpts from an email exchange between Embassy Baghdad foreign service officers Richard Albright and Alfred Anzaldua that can be found on pp. 58 and 60 of a batch of DoS documents released in response to an ACLU FOIA request. See https://www.aclu.org/files/pdfs/humanrights/irap_foia_release_state_95-219.pdf.

48. CWC 2011a.

49. Warren 2012.

50. Raustiala 2009; Potts 2017.

51. Grimmer 2013, 137.

52. Eichler 2014.

8. ACTIVISM

1. Stillman 2011.

2. Greene 2009.

3. CIA 1966b.

4. CIA 1966a.

5. Stanton 2003, 215.

6. Haberman 1986; J. Tyson 1986; Fineman 1986.

7. Vandiver 2013.

8. Holmes 2014.

9. Details in this paragraph come from a March 2015 interview with Endaya. On the Taji strike, see also Lee-Brago 2005.

10. Harris 2006.

11. Paley 2008.

12. A copy of ICAO's regulations concerning RFFS categories can be downloaded from https://www.bazl.admin.ch/dam/bazl/fr/dokumente/Fachleute/Flugplaetze/ICAO/icao _doc_9137_airportservicesmanualpart1withnoticeforusers.pdf.download.pdf/icao_doc _9137_airportservicesmanualpart1withnoticeforusers.pdf.

13. Hirschman 1970.

14. A Word document detailing these regulations can be downloaded at http://www .acq.osd.mil/dpap/policy/policyvault/Class_Deviation_2014-O0018_Attachment.docx.

15. For more on labor exploitation, trafficking, and the *kafala* system, see Ali 2010; Gardner 2010. In recent years several Gulf states have announced reforms to existing *kafala* labor practices, though human rights activists claim these formal changes have done little to improve labor migrants' conditions in practice. See, for example, Human Rights Watch's (2012) critique of Bahrain's 2009 reforms.

16. A Serka contract with this language can be found in a 2006 Army investigation of the company. See Harris 2006, exhibit 11. Examples of Najlaa and Kulak contracts were provided by Stillman as supplemental material accompanying her 2011 article. Though the online links no longer work the author has retained copies of the contracts.

17. This description comes from Card's website. See http://www.cardindustriesinc.com /about_us.php.

18. Boyles 2006. A text copy of this memo can be found at https://2001-2009.state.gov /g/tip/rls/other/2006/107279.htm.

19. See Streitfeld 2014. Court proceedings for the Silicon Valley case can be found at http://www.cand.uscourts.gov/lhk/hightechemployee. In 2015 the technology companies agreed to a $415 million settlement to compensate former employees harmed by their anti-poaching conspiracy. See Whitney 2015.

20. The above details are taken from two appeals of the original judgments, which were denied by the Department of Labor and Employment in 2013. Copies on file with author.

9. RELATIONS

1. U.S. Army Audit Agency 2013, 4.

2. "Agility," 2009.

3. See, for example, Higate 2012b; Chisholm 2014a, 2014b. Coburn (2018) provides a fascinating account of the recruitment, training, marketing, and afterlives of Gurkhas in Nepal (see, especially chaps. 6–8).

4. Quoted in Greene 2009, 47.

5. Quoted in Greene 2009, 49.

6. Greene 2009, 63.

7. Greene 2009, 64–67.

8. Greene 2009, 127.

9. Quoted in Greene 2009, 48.

10. Goran frames the issue as a lack of encounters with racial "others." But as Catherine Baker (2018) argues, it is a mistake to assume that the centrality of ethnic categories in Bosnia and other countries that constituted the former Yugoslavia means the region has existed "outside" of the politics of race and racialized imaginations, which are also deeply embedded in popular consciousness.

11. Sullivan 2010.

12. Crawford's website went dark in 2017, four years after she quit posting regularly, but her posts can still be found through the Wayback Machine. See p. 3 of the comments at https://web.archive.org/web/20111010084733/http://mssparky.com/2009/10/fluors -locap-iv-offer-for-kbr-employees-in-afghanistan/comment-page-3/#comments.

13. Ver 2008.

14. Guevarra 2009, 4; italics in original.

15. For a detailed analysis of this phenomenon in the cruise industry, see Terry 2014.

16. Coburn's (2018, 296) research on contractors in Afghanistan also suggests this is the case.

17. See, for example, Artwick 2013; Shor et al. 2015; Jia et al. 2016.

18. Eichler 2013, 312. For more on military contracting and masculinity, see Higate 2012a, 2012b; Stachowitsch 2013; Chisholm 2014a, 2014b, 2017; Stachowitsch 2015; Joachim and Schneiker 2015; Chisholm and Stachowitsch 2017.

19. Chisholm and Stachowitsch 2017, 378. Even the rare exceptions that examine logistics labor utilize the frame of masculinity and assume the absence of female workers. Isabelle Barker (2009), for instance, claims that the performance of "effeminately" coded "reproductive labor"—such as dining, billeting and laundry services—by poor migrant men from South and Southeast Asian countries reinforces an aggressive, masculine image of military service among American troops. While Barker's argument contains a kernel of truth, she was apparently unaware of the fact that women from Asia, Africa, and Southeast Europe are also involved in reproductive work, especially jobs related to laundry and billeting (this is likely due to the fact that Barker did not actually interview any TCN workers, but relied on news reportage and other secondary sources to construct her argument). Moreover, reproductive labor is but a small slice of the broad range of logistics work performed by both male and female TCNs.

20. Stoler 2001, 829. Stoler's work has been particularly influential in the growth of such studies. See, especially, Stoler 2002.

21. According to Ailyn, a Serka employee interviewed by Lee Wang for her documentary *Someone Else's War* (2006), this policy change occurred in 2006. Those I interviewed suggested this happened in 2008.

22. For more on this, see Vine 2015, chap. 10.

23. Zoroya 2012; Sadler et al. 2017.

24. Stillman 2011. For a legal analysis of the problem, see Snell 2011.

10. HOME

1. Moore 2017.

2. Skokić 2008; Huremović 2010; Husović 2010; Pargan 2010.

3. Srdjan was working for an off-base contractor at the time, and living in a private compound (actually a large guarded house) in Kabul, an arrangement that has been more common in Afghanistan than Iraq. For more on the lives of off-base contractors in Afghanistan, see Coburn 2018.

4. Commission on Filipinos Overseas 2013.

5. Guevarra 2010. For more on the deployment of *bagong bayani* as a political discourse, see Encinas-Franco 2013.

6. Quoted in Guevarra 2010, 52–53.

7. Quoted in Guevarra 2010, 55.

8. Chisholm 2014a.

9. Kelly and Lusis 2006, 840.

11. EMPIRE'S LABOR

Epigraph: Josh Begley, "How do you measure a military footprint?," http://empire.is /about.

1. Callimachi et al. 2018.

2. Department of the Air Force 2016.

3. Trevithick 2017.

4. Moore 2016c.

5. See https://www.fluor.com/careers/logcap-iv-opportunities.

6. Boot 2003, 42.

7. Gregory 2011.

8. Ryan 2011.

9. Keen 2006; Bacevich 2018; Filkins 2008. The "long war" was the favored nomenclature of neoconservative intellectuals that supported the Bush administration's "war on terror" policies that also had significant influence in military circles in the 2000s. See Bacevich 2007.

10. G Mann et al 2017, 269.

11. M. Shaw 2005. Recently Andreas Krieg and Jean-Marc Rickli (2018) have proposed the concept of "surrogate warfare" as a way of extending Shaw's original framework. On the "robotic revolution," see Singer 2009.

12. M. Shaw 2005, 71. See also M. Shaw 2002. For an earlier critique of the pursuit of "riskless" war and humanitarian interventions, see Kahn 1999.

13. Roderick 2010; Sauer and Schornig 2012. See also Chamayou (2013) on drones, risk and "combatant immunity."

14. Enloe 2010, vii.

15. For more on how contracting is impacting military operations and U.S. foreign policy, see Avant and de Nevers 2011.

16. This, and the following two paragraphs, draw on arguments made in Moore 2018.

17. I. Shaw 2016, 10 and 38.

18. Elish 2017, 1104.

19. Obama 2015.

20. Dodwell 2016.

21. One of the most vociferous hawks, John Bolton, is now President Trump's national security advisor.

22. I. Shaw 2016, 133.

23. DoD 2015.

24. See Peters, Schwartz, and Kapp 2016, 4.

25. Barkawi 2006, 74–75. For more on this, see Barkawi 2017.

Bibliography

301st Area Support Group (General Support Unit) Garrison Command. 2005. *LSA Anaconda master plan*. September. Copy on file with author.

Adhikary, Dhruba, and David Rohde. 2004. "Nepalese attack a mosque and Muslims in Katmandu." *New York Times*, September 2. http://www.nytimes.com/2004/09/02 /international/asia/02nepal.html?_r=0.

Agarib, Amira. 2015. "Dubai cops unravel 8-year operation Pink Panther." *Khaleej Times*, October 20. http://www.khaleejtimes.com/nation/crime/dubai-cops-unravel-8-year -operation-pink-panther.

"Agility says DynCorp ditches unit, mulling options." 2009. *ArabFinance*, December 12. https://www.arabfinance.com/2015/Pages/news/newsdetails.aspx?Id=157548 &lang=en.

"Agreement between the government of the United States of America and the government of the Republic of the Philippines relating to the recruitment and employment of Philippine citizens by the United States military forces and contractors of military and civilian agencies of the United States government in certain areas of the Pacific and Southeast Asia." Signed in Manila, December 28, 1968. Philippine Overseas Employment Administration. Copy on file with author.

Alderton, Tony, Michael Bloor, Erol Kahveci, Tony Lane, Helen Sampson, Michelle Thomas, Nik Winchester, Bin Wu, and Minghua Zhao. 2004. *The global seafarer: Living and working conditions in a globalized industry*. Geneva: International Labor Organization.

Ali, Syed. 2010. *Dubai: Gilded cage*. New Haven, CT: Yale University Press.

American Civil Liberties Union and Allard K. Lowenstein International Human Rights Clinic. 2012. *Victims of complacency: The ongoing trafficking and abuse of third country nationals by U.S. government contractors*. https://www.aclu.org/files/assets/hrp _traffickingreport_web_0.pdf.

Amnesty International. 2016. *The ugly side of the beautiful game: Exploitation of migrant workers on a Qatar 2022 World Cup site*. https://www.amnesty.org/download /Documents/MDE2235482016ENGLISH.PDF.

Anderson, Steven. 2011. "Save energy, save our troops." *New York Times*, January 2. http:// www.nytimes.com/2011/01/13/opinion/13anderson.html?_r=1.

Andzans, Maris. 2013. "The Northern Distribution Network and its implications for Latvia." In *Northern Distribution Network: Redefining partnerships within NATO and beyond*, edited by Diana Potjomkina and Andris Spruds, 9–30. Riga: Latvian Institute of International Affairs.

Appel, Hannah. 2012. "Offshore work: Oil, modularity, and the how of capitalism in Equatorial Guinea." *American Ethnologist* 39(4): 692–709.

Armakolas, Ioannis. 2011. "The 'paradox' of Tuzla city: Explaining non-nationalist local politics during the Bosnian war." *Europe-Asia Studies* 63(2): 229–61.

Arnold, Wayne. 2003. "The postwar invasion of Iraq." *New York Times*, April 9. http://www .nytimes.com/2003/04/09/business/the-postwar-invasion-of-iraq.html.

Arsenijević, Damir. 2014. *Unbribable Bosnia and Herzegovina: The fight for the commons*. Baden-Baden, Germany: Nomos.

Artwick, Claudette. 2013. "News sourcing and gender on Twitter." *Journalism* 15(8): 1111–27.

Attewell, Wesley. 2017. "'The planet that rules our destiny': Alternative development and environmental power in Afghanistan." *Environment and Planning D: Society and Space* 35(2): 339–59.

Avant, Deborah. 2005. *The market for force: The consequences of privatizing security.* Cambridge: Cambridge University Press.

Avant, Deborah, and Rennee de Nevers. 2011. "Military contracting and the American way of war." *Daedalus* 140(3): 88–99.

Avant, Deborah, and Lee Sigelman. 2010. "Private security and democracy: Lessons from the US in Iraq." *Security Studies* 19(2): 230–65.

Bacevich, Andrew, ed. 2007. *The long war: A new history of U.S. national security policy since World War II.* New York: Colombia University Press.

———. 2018. "Infinite war: The gravy train rolls on." *TomDispatch.com*, June 7. http://www .tomdispatch.com/blog/176433/tomgram%3A_andrew_bacevich%2C_not_so _great_wars%2C_theirs_and_ours.

Bachmann, Jan. 2010. "'Kick down the door, clean up the mess and rebuild the house'— the Africa Command and the transformation of the U.S. military." *Geopolitics* 15(3): 564–85.

Baker, Catherine. 2012. "Prosperity without security: The precarity of interpreters in post-socialist, postconflict Bosnia and Herzegovina." *Slavic Review* 71(4): 849–72.

———. 2014. "The local workforce of international intervention in the Yugoslav successor states: 'Precariat' or 'projectariat'? Towards and agenda for future research." *International Peacekeeping* 21(1): 91–106.

———. 2018. *Race and the Yugoslav region: Postsocialist, post-conflict, postcolonial?* Manchester: Manchester University Press.

Baker, Norman. 1971. *Government and contractors: The British Treasury and war supplies, 1775–1783.* London: Athlone Press.

Baldoz, Rick. 2011. *The third Asiatic invasion: Migration and empire in Filipino America: 1898–1946.* New York: New York University Press.

Bandjunis, Vytautas. 2001. *Diego Garcia: Creation of the Indian Ocean base.* Lincoln, NE: Writers Showcase.

Bankoff, Greg. 2005. "Wants, wages, and workers: Laboring in the American Philippines, 1899–1908." *Pacific Historical Review* 74(1): 59–86.

Barkawi, Tarak. 2006. *Globalization and war.* Lanham, MD: Rowman and Littlefield.

———. 2017. *Soldiers of empire: Indian and British armies in WWII.* Cambridge: Cambridge University Press.

Barker, Isabelle. 2009. "(Re)producing American soldiers in an age of empire." *Politics & Gender* 5(2): 211–35.

Battistella, Graziano. 1999. "Philippine migration policy: Dilemmas of a crisis." *Sojourn: Journal of Social Issues in Southeast Asia* 14(1): 229–48.

Beasley, Betsy. 2019. "The strange career of Donald Rumsfeld: Military logistics and the routes from Vietnam to Iraq." *Radical History Review* 133: 56–77.

Belanger, Pierre, and Alexander Arroyo. 2012. "Logistics islands: The global supply archipelago and the topologics of defense." *Prism: A Journal of the Center of Complex Operations* 3(4): 55–75.

———. 2016. *Ecologies of power: Countermapping the logistical landscapes and military geographies of the U.S. Department of Defense.* Cambridge, MA: MIT Press.

Belasco, Amy. 2009. *Troop levels in the Afghan and Iraq wars, FY 2001–FY2012: Cost and other potential issues.* Congressional Research Service. July 2. https://fas.org/sgp/crs /natsec/R40682.pdf.

Bell, Thomas. 2004. "Three dead as Nepalis riot over Iraq deaths." *Telegraph*, September 2. http://www.telegraph.co.uk/news/worldnews/middleeast/iraq/1470820/Three -dead-as-Nepalis-riot-over-Iraq-deaths.html.

Bender, Daniel, and Jana Lipman. 2015. "Introduction." In *Making the empire work: Labor and United States imperialism*, edited by Daniel Bender and Jana Lipman, 1–34. New York: New York University Press.

Berry, William. 1980. "American military bases in the Philippines, base negotiations, and Philippine-American relations: Past, present, and future." PhD diss., Cornell University.

Bianco, Anthony, and Stephanie A. Forest. 2003. "Outsourcing war." *Bloomberg*, September 14. http://www.bloomberg.com/news/articles/2003-09-14/outsourcing-war.

Birks, J. S., I. J. Seccombe, and C. A. Sinclair. 1988. "Labor migration in the Arab Gulf States: Patterns, trends and prospects." *International Migration* 26(3): 267–86.

Black, Samuel, and Anjali Kamat. 2014. "After 12 years of war, labor abuses rampant on US bases in Afghanistan." *Al Jazeera America*, March 7. http://america.aljazeera.com /articles/2014/3/7/after-12-years-ofwarlaborabusesrampantonusbasesinafghanist an.html.

Bloor, Michael, and Helen Sampson. 2009. "Regulatory enforcement of labour standards in an outsourcing globalized industry: The case of the shipping industry." *Work, Employment and Society* 23(4): 711–26.

Bonacich, Edna, and Jake Wilson. 2008. *Getting the goods: Ports, labor, and the logistics revolution*. Ithaca, NY: Cornell University Press.

Boot, Max. 2003. "The new American way of war." *Foreign Affairs* 82(4): 41–58.

Boyd, E. B. 2015. "Getting out of Afghanistan." *Fast Company*, January 28. https://www .fastcompany.com/3041147/getting-out-of-afghanistan.

Boyles, Robert K. 2006. "Memorandum for all contractors. Subject: Withholding of passports, trafficking of persons." Joint Contracting Command Iraq/Afghanistan. April 19. Copy on file with author.

Briody, Dan. 2004. *The Halliburton agenda: The politics of oil and money*. Hoboken, NJ: John Wiley and Sons.

Brkić, Denis. 2012. "Lukavčane potraga za boljim životom vodi u Afganistan" [Lukavac residents' search for a better life leads to Afghanistan]. *Sodalive.ba*, January 31. www .sodalive.ba/aktuelnosti/lukavcane-potraga-za-boljim-zivotom-vodi-u-afganistan.

Brown, Keith. 2010. "From the Balkans to Baghdad (via Baltimore): Labor migration and the routes of empire." *Slavic Review* 69(4): 816–34.

Brown, Robbie. 2009. "U.S. says Kuwaiti company overbilled it by millions for troops' food." *New York Times*, November 16. http://www.nytimes.com/2009/11/17/world /middleeast/17fraud.html?_r=0.

Bruneau, Thomas C. 2011. *Patriots for profit: Contractors and the military in U.S. national security*. Stanford, CA: Stanford University Press.

Bureau of Insular Affairs. 1905. *Sixth annual report of the Philippine Commission, 1905 (Part 3)*. Washington, DC: War Department.

Cahlink, George A. 2002. "Army of contractors." *Government Executive* 34(2): 43–45.

——. 2003. "Send in the contractors." *Air Force Magazine*, January. http://www.airforcemag .com/magazinearchive/documents/2003/january%202003/0103contract.pdf.

Callimachi, Rukmini, Helene Cooper, Eric Schmitt, Alan Blinder, and Thomas Gibbons-Neff. 2018. "'An endless war': Why 4 U.S. soldiers died in a remote African desert." *New York Times*, February 17. https://www.nytimes.com/interactive/2018/02/17 /world/africa/niger-ambush-american-soldiers.html.

Campbell, John. 2015. "Bad news in Burkina Faso." *Africa in Transition* (blog), September 17. http://blogs.cfr.org/campbell/2015/09/17/bad-news-in-burkina-faso.

Caro, Robert A. 1990. *Means of ascent.* New York: Vintage Books.

Carter, James M. 2008. *Inventing Vietnam: The United States and statebuilding, 1954–1968.* Cambridge: Cambridge University Press.

Castaneda, Dabet. 2004. "Barangay Iraq." *Bulatlat* 4(26), August 1–7. http://www.bulatlat .com/news/4-26/4-26-barangay.html.

Celoza, Albert. 1997. *Ferdinand Marcos and the Philippines: The political economy of authoritarianism.* Westport, CT: Praeger.

Central Intelligence Agency. 1966a. *Intelligence report: The situation in South Vietnam.* July 25. https://www.cia.gov/library/readingroom/docs/CIA-RDP79T00826A001 000010055-4.pdf.

——. 1966b. *Memorandum: The situation in Vietnam.* July 29. https://www.cia.gov/library /readingroom/docs/CIA-RDP79T00826A001100010001-2.pdf.

——. 1969a. *Intelligence report: The economic situation in South Vietnam.* June 2. https:// www.cia.gov/library/readingroom/docs/CIA-RDP82S00205R000200010035-4 .pdf.

——. 1969b. *Intelligence report: The economic situation in South Vietnam.* December 29. Copy on file with author.

Cha, Ariana. 2004. "Underclass of workers created in Iraq: Many foreign laborers receive inferior pay, food and shelter." *Washington Post,* July 1. http://www.washingtonpost .com/wp-dyn/articles/A19228-2004Jun30.html.

Chamayou, Gregoire. 2013. *A theory of the drone.* New York: New Press.

Chatterjee, Pratap. 2009. *Halliburton's army: How a well-connected Texas oil company revolutionized the way America makes war.* New York: Nation Books.

Chisholm, Amanda. 2014a. "Marketing the Gurkha security package: Colonial histories and neoliberal economies of private security." *Security Dialogue* 45(4): 349–72.

——. 2014b. "The silenced and indispensable: Gurkhas in private military companies." *International Feminist Journal of Politics* 16(1): 26–47.

——. 2017. "Clients, contractors, and everyday masculinities in global private security." *Critical Military Studies* 3(2): 120–41.

Chisholm, Amanda, and Saskia Stachowitsch. 2017. "Military markets, masculinities and the global political economy of the everyday: Understanding military outsourcing as gendered and racialized." In *The Palgrave handbook of gender and the military,* edited by Rachel Woodward and Claire Duncanson, 371–96. London: Palgrave.

Choy, Catherine Ceniza. 2003. *Empire of care: Nursing and migration in Filipino-American history.* Durham, NC: Duke University Press.

Chua, Charmaine, Martin Danyluk, Deborah Cowen, and Laleh Khalili. 2018. "Introduction: Turbulent circulation: Building a critical engagement with logistics." *Environment and Planning D: Society and Space* 36(4): 617–29.

Chung, Patrick. 2019. "From Korea to Vietnam: Local labor, multinational capital, and the evolution of U.S. military logistics, 1950-97." *Radical History Review* 133: 31–55.

Cleveland, Bradley A. 2008. "The last shall be first: The use of localized socio-economic policies in contingency contracting operations." *Military Law Review* 197(Fall): 103–44.

Coburn, Noah. 2018. *Under contract: The invisible workers of America's global war.* Stanford, CA: Stanford University Press.

Cohen, William S. 1997. "Time has come to leap into the future." *Defense Issues* 12(19): 1–4.

Commission on Filipinos Overseas. 2013. *Stock estimate of overseas Filipinos (as of December 2013).* http://www.cfo.gov.ph/images/stories/pdf/StockEstimate2013.pdf.

Commission on Wartime Contracting in Iraq and Afghanistan. 2011a. *At what risk? Correcting over-reliance on contractors in contingency operations.* February 24. http://

cybercemetery.unt.edu/archive/cwc/20110929221313/http://www.wartimecontract
ing.gov/docs/CWC_InterimReport2-lowres.pdf.

———. 2011b. *Transforming wartime contracting: Controlling costs, reducing risks.* August.
http://cybercemetery.unt.edu/archive/cwc/20110929213820/http://www
.wartimecontracting.gov/docs/CWC_FinalReport-lowres.pdf.

Conde, Carlos. 2004. "Filipinos still seek work in Iraq despite danger and ban." *New York
Times,* September 3. http://www.nytimes.com/2004/09/03/world/asia/filipinos-still
-seek-work-in-iraq-despite-danger-and-ban.html.

Conner, Jon Michael. 2015. "ASC's LOGCAP plays crucial role in fight against Ebola." U.S.
Army. http://www.army.mil/article/148303/ASC_s_LOGCAP_plays_crucial_role_in
_fight_against_Ebola.

Cooley, Alexander. 2008. *Base politics: Democratic change and the U.S. military overseas.*
Ithaca, NY: Cornell University Press.

———. 2012. *Great games local rules: The new great power contest in Central Asia.* Ithaca,
NY: Cornell University Press.

Cornella, Al, Lewis E. Curtis III, Anthony Less, Keith Martin, H. G. (Pete) Taylor, and
James A. Thomson. 2005a. *Report to the President and Congress.* Overseas Basing
Commission. May 9. http://fas.org/irp/agency/dod/obc.pdf.

———. 2005b. *Report to the President and Congress.* Overseas Basing Commission. August 15.
http://govinfo.library.unt.edu/osbc/documents/OBC%20Final%20Report%20
August%2015%202005.pdf.

Corrick, David. 2012. "The new spice route for Africa." *Army Sustainment,* March–April:
46–47. http://www.almc.army.mil/alog/issues/MarApril12/pdf/Mar-Apr2012.pdf.

Cowen, Deborah. 2013. *The deadly life of logistics: Mapping violence in global trade.* Minneapolis, MN: University of Minnesota Press.

Crampton, Jeremy W., Susan R. Roberts, and Ate Poorthuis. 2014. "The new political
economy of geographical intelligence." *Annals of the Association of American Geographers* 104(1): 196–214.

Cusumano, Eugenio. 2018. "Resilience for hire? NATO contractor support in Afghanistan
examined." In *A civil-military response to hybrid threats,* edited by Eugenio Cusumano and Marian Corbe, 101–22. Cham, Switzerland: Palgrave.

Dacanay, Barbara Mae. 2004. "Deployment of drivers to Iraq banned." *Gulf News,* May 16.
http://gulfnews.com/news/asia/philippines/deployment-of-drivers-to-iraq
-banned-1.322481.

Daly, John. 2015. "Central Asia will miss the Northern Distribution Network." *Silk Road
Reporters,* June 19. https://web.archive.org/web/20160703063959/http://www.silk
roadreporters.com/2015/06/19/central-asia-will-miss-the-northern-distribution
-network.

Dao, James. 2002. "Rumsfeld resists calls to build up military forces." *New York Times,*
April 19. http://www.nytimes.com/2002/04/19/us/rumsfeld-resists-calls-to-build
-up-military-forces.html.

Darwin, John. 2008. *After Tamerlane: The global history of empire since 1405.* New York:
Bloomsbury Press.

Davies, Sara. 2008. "Securitizing infectious diseases." *International Affairs* 84(2): 295–313.

Delo, Michael D. 1992. *Peddlers and post traders: The Army sutler on the frontier.* Salt Lake
City: University of Utah Press.

Defense Science Board Task Force. 1996. *Report of the Defense Science Board Task Force on
privatization and outsourcing.* Washington, DC: Office of the Under Secretary of Defense for Acquisition and Technology.

———. 2014. *Report of the Defense Science Board Task Force on contractor logistics in support
of contingency operations.* Washington, DC: Office of the Under Secretary of Defense

for Acquisition, Technology and Logistics. http://www.acq.osd.mil/dsb/reports /CONLOG_Final_Report_17Jun14.pdf.

Department of the Air Force. 2016. *Statement of work for replace legacy latrines and showers Niamey, Niger*. July 28. https://www.fbo.gov/utils/view?id=9e7ae919dd7d5f0cb 32e61e2701daba6.

Department of the Army. 1985. *Army regulation 700-137: Logistics Civilian Augmentation Program (LOGCAP)*. December 16. http://www.aschq.army.mil/gc/files/AR700-137 .pdf.

———. 2016. *Multi-service tactics, techniques, and procedures for operational contract support. (ATP 4-10)*. February. http://armypubs.army.mil/doctrine/DR_pubs/dr_a/pdf/attp4 _10.pdf.

Department of Defense. 1992. *Final report to Congress: Conduct of the Persian Gulf War*. https://apps.dtic.mil/dtic/tr/fulltext/u2/a249270.pdf.

———. 2015. *Contractor support of U.S. operations in the USCENTCOM area of responsibility*. https://www.acq.osd.mil/log/ps/.CENTCOM_reports.html/5A_January2015.pdf.

Department of Defense, Office of Inspector General. 2015. *Army needs to improve contract oversight for the Logistics Civil Augmentation Program's task orders*. October 28. https://media.defense.gov/2015/Oct/28/2001714178/-1/-1/1/DODIG-2016-004 .pdf.

Department of Labor, Office of Workers' Compensation Program. n.d. "Office of Workers' Compensation Programs (OWCP)." https://www.dol.gov/owcp/dlhwc/dbaallnation .htm.

Department of State. *Trafficking in Persons Report 2011*. https://www.state.gov/j/tip/rls /tiprpt/2011/index.htm.

Deptula, David. 2010. *The way ahead: Remotely piloted aircraft in the United States Air Force*. January 25. https://web.archive.org/web/20140715005916/http://www.daytonregion .com/pdf/UAV_Rountable_5.pdf.

De Simone, Anthony, and Norm Gauthier. 2003. "The Inland Petroleum Distribution System in Kuwait and Iraq." *Engineer: The Professional Bulletin of Army Engineers* (July–September): 13–18. http://www.wood.army.mil/engrmag/PDFs%20for%20 Jul-Sept%2003/De%20Simone.pdf.

De Vera, Ben O. 2017. "OFW remittances hit record high in 2016." *Philippine Daily Inquirer*, February 16. http://business.inquirer.net/224635/ofw-remittances-hit-record -high-2016.

DeYoung, Karen. 2011. "U.S. trucking funds reach Taliban, military-led investigation concludes." *Washington Post*. July 24. https://www.washingtonpost.com/world /national-security/us-trucking-funds-reach-taliban-military-led-investigation -concludes/2011/07/22/gIQAmMDUXI_story.html?utm_term=.79d1de820189.

Dickey S. L. 2013. *A "new normal": The U.S. Marine Corps' approach to meeting evolving global security requirements*. May 10. http://www.afcea.org/events/jwc/13/documents /AFCEAUSNI10May131Dickey.pdf.

Directorate of Logistics. 2015. "Operation United Assistance: Logistics partnership success." United States Africa Command. January 12. http://www.africom.mil/NewsByCate gory/Article/25102/operation-united-assistance-logistics-partnership-success.

Dodd, Joseph. 1968. *Criminal jurisdiction under the United States-Philippine Military Bases Agreement*. The Hague: Martinus Nijhoff.

Dodwell, Brian. 2016. "A view from the CT foxhole: Brigadier General Donald C. Bolduc, commander, Special Operations Command Africa." *CTC Sentinel*, May 25. https:// www.ctc.usma.edu/posts/a-view-from-the-ct-foxhole-brigadier-general-donald-c -bolduc-commander-special-operations-command-africa.

Duaqui, Yellowbell. 2013. "From Marcos to Aquino governments: State sponsorship of di-aspora philanthropy." *Philippine Social Sciences Review* 65(2): 74–99.

Dunigan, Molly. 2011. *Victory for hire: Private security companies' impact on military ef-fectiveness.* Stanford, CA: Stanford University Press.

Dunn, Carroll H. 1972. *Base development in South Vietnam: 1965–1970.* Washington, DC: Department of the Army.

Duroni, Lance. 2013. "Kuwaiti co. sues KBR to secure $12.6M arbitration win. *Law360,* June 21. https://www.law360.com/articles/451938.

Eakin, Emily. 2002. "All roads lead to DC." *New York Times,* April 1. http://www.nytimes.com/learning/teachers/featured_articles/20020401monday.html.

Eckert, Amy E. 2016. *Outsourcing war: The just war tradition in the age of military privati-zation.* Ithaca, NY: Cornell University Press.

Eichler, Maya. 2013. "Gender and the privatization of security: Neoliberal transformation of the militarized gender order." *Critical Studies of Security* 1(3): 311–25.

——. 2014. "Citizenship and the contracting out of military work: From national con-scription to globalized recruitment." *Citizenship Studies* 18(6–7): 600–14.

——, ed. 2015. *Gender and private security in global politics.* New York: Oxford University Press.

Elish, M. C. 2017. "Remote split: A history of drone operations and the distributed labor of war." *Science, Technology & Human Values* 42(6): 1100–1131.

Ellison, Keith. 2014. *Memorandum and order: Ramchandra Adhikari et al vs. Daoud & Part-ners et al.* United States District Court, Southern District of Texas, Houston Divi-sion. January 15. https://www.gpo.gov/fdsys/pkg/USCOURTS-txsd-4_09-cv-01237/pdf/USCOURTS-txsd-4_09-cv-01237-4.pdf.

Embassy Baghdad. 2009. "KBR subcontractor suspected of TIP violations." Department of State cable (Unclassified/For Official Use Only). January 7. https://wikileaks.org/plusd/cables/09BAGHDAD35_a.html.

Embassy Islamabad. 2009. "Action on Pak-Afghan transit trade negotiations." Department of State cable (Unclassified/For Official Use Only). May 1. https://wikileaks.org/plusd/cables/09ISLAMABAD930_a.html.

Embassy Kathmandu. 2004. "Nepal: Sitrep 2: Quiet on Sept. 2: National day of mourn-ing." Department of State cable (Confidential). September 2. https://search.wikileaks.org/plusd/cables/04KATHMANDU1759_a.html.

Embassy Kuwait. 2003. "Kuwait: 2004 annual report to Congress on allied contributions to the common defense." Department of State cable (Secret). December 23. https://search.wikileaks.org/plusd/cables/03KUWAIT5788_a.html.

——. 2004a. "Outreach to contractors on security and labor issues in Iraq." Department of State cable (Sensitive, Kuwait 1598). May 19. Copy on file with author.

——. 2004b. "Indian workers in Iraq: Indian Embassy requests follow up on CFLCC pro-posals for workers' rights." Department of State cable (Sensitive, Kuwait 2006). June 29. Copy on file with author.

——. 2004c. "India and the Philippines impose ban on citizens' travel to Iraq." State Department cable (Confidential). August 4. https://wikileaks.org/cable/2004/08/04KUWAIT2425.html.

——. 2004d. "Iraq travel ban: U.S. military presses India and Philippines to exempt na-tionals under military escort." Department of State cable (Confidential). August 7. https://wikileaks.org/cable/2004/08/04KUWAIT2496.html.

——. 2004e. "Iraq travel ban: Nepal enforces ban on its nationals, U.S. military provides welfare assurances." Department of State cable (Secret). September 9. https://wikileaks.org/cable/2004/09/04KUWAIT3033.html.

——. 2004f. "Kuwait MFA says GOK will not enforce ban on foreign nationals entering Iraq." Department of State cable (Confidential). September 21. https://wikileaks.org/cable/2004/09/04KUWAIT3262.html.

——. 2005. "CFLCC/MOI sign border MOU." Department of State cable (Unclassified). July 6. https://wikileaks.org/plusd/cables/05KUWAIT3037_a.html.

——. 2006. "Kuwait: 2005/2006 report to Congress on allied contributions to the common defense." State Department cable (Secret). February 6. https://search.wikileaks.org/plusd/cables/06KUWAIT407_a.html.

——. 2007a. "Reconnaissance visit to new Iraq-Kuwait border crossing." Department of State cable (Confidential). 9 January 9. https://wikileaks.org/plusd/cables/07KUWAIT33_a.html.

——. 2007b. "Visit to three Kuwait-Iraq border facilities." Department of State cable (Confidential). January 31. https://wikileaks.org/plusd/cables/07KUWAIT147_a.html.

——. 2007c. "New Kuwait-Iraq border crossing to open May 15." Department of State cable (Confidential). May 9. https://wikileaks.org/plusd/cables/07KUWAIT730_a.html.

——. 2007d. "Scenesetter for Secretary of Defense Gates' visit to Kuwait." Department of State cable (Secret). July 26. https://wikileaks.org/plusd/cables/07KUWAIT1170_a.html.

——. 2009. "A big footprint in the sand: The U.S. presence in Kuwait." Department of State cable (Secret). October 29. https://wikileaks.org/plusd/cables/09KUWAIT1036_a.html.

Embassy Manila. 2004a. "Working to lift the travel ban to Iraq." Department of State cable (Confidential, Manila 3893). Copy on file with author.

——. 2004b. "Ambassador presses Philippine vice president on Iraq travel ban." Department of State cable (Confidential, Manila 4020). Copy on file with author.

Embassy New Delhi. 2004. "Mission debunks media reports of abuse of Indian workers by U.S. Army." Department of State cable (Confidential, New Delhi 2766). May 6. Copy on file with author.

Embassy Tashkent. 2009. "Uzbekistan scenesetter for General Petraeus." State Department cable (Confidential). February 11. https://wikileaks.org/plusd/cables/09TASHKENT162_a.html.

Encinas-Franco, Jean. 2013. "The language of labor export in political discourse: 'Modern-day heroism' and constructions of overseas Filipino workers (OFWs)." *Philippine Political Science Journal* 34(1): 97–112.

Englehart, Tom. 2009. "Iraq as a Pentagon construction site." In *The bases of empire: The global struggle against U.S. military posts*, edited by Catherine Lutz, 131–44. New York: New York University Press.

Enloe, Cynthia. 2010. "Foreword." In *Militarized currents: Toward a decolonized future in Asia and the Pacific*, edited by Setsu Shigematsu and Keith Camacho, vii–ix. Minneapolis, MN: University of Minnesota Press.

Erbel, Mark, and Christopher Kinsey. 2018. "Think again—supplying war: Reappraising military logistics and its centrality to strategy and war." *Journal of Strategic Studies* 41(4): 519–44.

Ettlinger, Nancy. 2007. "Precarity unbound." *Alternatives* 32(3): 319–40.

Farrand, Dale. 2006. "The 'revolution in military logistics': Is it enough?" Master's thesis, U.S. Army Command and General Staff College. http://www.dtic.mil/dtic/tr/fulltext/u2/a463836.pdf.

FDCH E-Media. 2005. "Transcript: Rumsfeld testifies before House panel." *Washington Post*, February 16. http://www.washingtonpost.com/wp-dyn/articles/A29752-2005Feb16.html.

Ferris, Stephen, and David Keithly. 2001. "Outsourcing the sinews of war: Contractor logistics." *Military Review* 81(5): 72–83.

Fialka, John J. 1990. "U.S. troops replace building boom as source of profit for Saudi businesses." *Wall Street Journal*, December 27.

Filkins, Dexter. 2008. *The forever war*. New York: Alfred Knopf.

Fineman, Mark. 1986. "U.S. bases in Philippines reopened as workers end strike." *Los Angeles Times*, April 3. http://articles.latimes.com/1986-04-03/news/mn-2650_1_military -base.

Finer, Jonathan. 2008. "Holstering the hired guns: New accountability measures for private security contractors." *Yale Journal of International Law* 33: 259–65.

Flores, Alfred Peredo. 2015. "'No walk in the park': US empire and the racialization of civilian military labor in Guam, 1944–1962." *American Quarterly* 67(3): 813–35.

Fontaine, Richard, and Jon Nagl. 2010. *Contracting in conflicts: The path to reform*. Center for New American Security. http://www.cnas.org/files/documents/publications /CNAS_Contracting%20in%20Conflicts_Fontaine%20Nagl.pdf.

Fregonese, Sara, and Adam Ramadan. 2015. "Hotel geopolitics: A research agenda." *Geopolitics* 20(4): 793–813.

Fujita-Rony, Dorothy. 2015. "Empire and the moving body: Fermin Tobera, military California, and rural space." In *Making the empire work: Labor and United States imperialism*, edited by Daniel Bender and Jana Lipman, 208–26. New York: New York University Press.

Gallaher, Carolyn. 2012. "Risk and private military work." *Antipode* 44(3): 783–805.

Gallay, David R., and Charles L. Horne. 1996. *LOGCAP support in Operation Joint Endeavor: A review and analysis*. McLean, VA: Logistic Management Institute. Cited in Maria J. Dowling and Vincent J. Fleck. 1999. *Feasibility of a joint engineering and logistics contract*. Alabama, GA: Air University Press.

Gardner, Andrew. 2010. *City of strangers: Gulf migration and the Indian community in Bahrain*. Ithaca, NY: Cornell University Press.

Gilbert, Bonita. 2012. *Building for war: The epic saga of the civilian contractors and Marines of Wake Island in World War II*. Oxford: Casemate.

Gisick, Michael. 2010. "U.S. base projects continue in Iraq despite plans to leave." *Stars and Stripes*, June 1. http://www.stripes.com/news/u-s-base-projects-continue-in -iraq-despite-plans-to-leave-1.105237.

Glassman, Jim, and Choi Young-Jin. 2014. "The *chaebol* and the US military-industrial complex: Cold War geopolitical economy and South Korean industrialization." *Environment and Planning A* 46(5): 1160–80.

Go, Julian. 2005. "Modes of rule in America's overseas empire: The Philippines, Puerto Rico, Guam, and Samoa." In *The Louisiana Purchase and American expansion, 1803–1898*, edited by Sanford Levinson and Bartholomew Sparrow, 209–29. Lanham, MD: Rowman and Littlefield.

——. 2007. "The provinciality of American empire: 'Liberal exceptionalism' and U.S. colonial rule, 1898–1912." *Comparative Studies in Society and History* 49(1): 74–108.

——. 2011. *Patterns of empire: The British and American empires, 1688 to the present*. New York: Cambridge University Press.

Gordan, Neil. 2014. "Court upholds 3-year contracting suspension." *Project on Government Oversight*, January 7. http://www.pogo.org/blog/2014/01/20140107-court -upholds-3-year-contracting-suspension.html.

Government Accounting Office. 1977. *Department of Defense should change pay setting for Filipino Nationals*. Comptroller General of the United States. October 7. http://www .gao.gov/assets/130/120297.pdf.

——. 1997. *Contingency operations: Opportunities to improve logistics civil augmentation program.* February 11. http://www.gao.gov/assets/230/223644.pdf.

Government of the Philippines. 1974. "Presidential Decree 442." https://web.archive.org /web/20161208121740/http://www.gov.ph:80/1974/05/01/presidential-decree-no -442-s-1974/.

Grace, J. Peter. 1984. *Burning money: The waste of your tax dollars.* New York: Macmillan.

Graham, Stephen. 2009. "Cities as battlespace: The new military urbanism." *City* 13(4): 383–402.

Grasso, Valerie B. 2008. *Defense contracting in Iraq: Issues and options for Congress.* Congressional Research Service. August 15. http://www.dtic.mil/dtic/tr/fulltext/u2 /a486297.pdf.

Greene, Julie. 2009. *The canal builders: Making America's empire at the Panama Canal.* New York: Penguin.

——. 2015. "The wages of empire: Capitalism, expansionism, and working-class formation." In *Making the empire work: Labor and United States imperialism,* edited by Daniel Bender and Jana Lipman, 35–58. New York: New York University Press.

——. 2016. "Builders of empire: Rewriting the labor and working-class history of Anglo-American global power." *Labor: Studies in Working-Class History of the Americas* 13(3–4): 1–10.

Gregory, Derek. 2006. "The black flag: Guantanamo Bay and the space of exception." *Geografiska Annaler: Series B, Human Geography* 88(4): 405–27.

——. 2010. "War and peace." *Transactions of the British Institute of Geographers* 35(2): 154–86.

——. 2011. "The everywhere war." *Geographical Journal* 177(3): 238–50.

——. 2012. "Supplying war in Afghanistan: The frictions of distance." *openDemocracy,* June 11. https://www.opendemocracy.net/derek-gregory/supplying-war-in-afghani stan-frictions-of-distance#_ednref24.

Greitens, Eric. 2011. *The heart and the fist: The education of a humanitarian, the making of a Navy SEAL.* New York: Houghton, Mifflin Harcourt.

Grimmer, Caitlyn. 2013. "Procuring protection: Using the False Claims Act to combat human trafficking by government contractors." *Public Contract Law Journal* 43(1): 127–44.

Guevarra, Anna Romina. 2010. *Marketing dreams, manufacturing heroes: The transnational labor brokering of Filipino workers.* New Brunswick, NJ: Rutgers University Press.

"Gulf countries employ highest number of foreign workers." 2016. *Daily Sabah,* January 27. https://www.dailysabah.com/business/2016/01/28/gulf-countries-employ-highest -number-of-foreign-workers.

Haberman, Clyde. 1986. "Filipino strikers picket U.S. bases." *New York Times,* March 23. http://www.nytimes.com/1986/03/23/world/filipino-strikers-picket-us-bases.html ?mcubz=3.

Harris, Brian. 2006. *CID report of investigation—final—0005-06-CID389-76408-8F5.* Department of the Army. August 19. Copy on file with author.

Higate, Paul. 2012a. "Drinking vodka from the 'butt-crack': Men, masculinities and fratriarchy in the private militarized security company." *International Feminist Journal of Politics* 14(4): 450–69.

——. 2012b. "Martial races and enforcement masculinities of the Global South: Weaponizing Fijian, Chilean, and Salvadoran postcoloniality in the mercenary sector." *Globalizations* 9(1): 35–52.

Hirsh, Max. 2017. "Emerging infrastructures of low-cost aviation in Southeast Asia." *Mobilities* 12(2): 259–76.

Hirschman, Albert. 1970. *Exit, voice, and loyalty: Responses to decline in firms, organizations, and states.* Cambridge, MA: Harvard University Press.

Holmes, Amy. 2014. *Social unrest and American military bases in Turkey and Germany since 1945.* New York: Cambridge University Press.

Human Rights Watch. 2009. *"The island of happiness": Exploitation of migrant workers on Saadiyat Island, Abu Dhabi.* https://www.hrw.org/sites/default/files/reports/uae0509 web_4.pdf.

———. 2012. *For a better life: Migrant worker abuse in Bahrain and the government reform agenda.* https://www.hrw.org/sites/default/files/reports/bahrain1012ForUpload.pdf.

Huremović, Elvir. 2010. "Bikovići ostali bez sina jedinica: U Afganistanu poginuo mladi Lukavčanin" [Bikovićes lose their only son: A young Lukavacan dies in Afghanistan]. *Dnevni Avaz,* March 16.

Husović, Edina. 2010. "Dženaza Almir Biković" [Funeral of Almir Biković]. *RTVLukavac. ba,* March 24. https://www.youtube.com/watch?v=-eEdAcNl_Fk.

Huston, James A. 1966. *The sinews of war: Army logistics, 1775–1953.* Washington, DC: Office of the Chief of Military History.

———. 1989. *Guns and butter, powder and rice: U.S. Army logistics in the Korean War.* Cranbury, NJ: Associated University Presses.

Hyndman, Jennifer. 2007. "Feminist geopolitics revisited: Body counts in Iraq." *Professional Geographer* 59(1): 35–46.

Ingram, Timothy. 1970. "The floating plantation." *Washington Monthly,* October: 17–20.

Isenberg, David. 2008. *Shadow force: Private security contractors in Iraq.* Westport, CT: Praeger Security International.

Isenberg, David, and Nick Schwellenbach. 2011. "Documents reveal details of alleged labor trafficking by KBR subcontractor: The Najlaa episode revisited." *Project on Government Oversight.* June 14. http://www.pogo.org/our-work/articles/2011/co-ht-2011 0614.html.

ISR Task Force, Requirements and Analysis Division. 2013. *ISR support to small footprint CT operations—Somalia/Yemen.* February. https://theintercept.com/document/2015 /10/15/small-footprint-operations-2-13.

Jackson, Justin. 2014. "The work of empire: The U.S. Army and the making of American colonialisms in Cuba and the Philippines, 1898–1913." PhD diss., Columbia University.

Jansen, Stef. 2006. "The privatisation of home and hope: Return, reforms and the foreign intervention in Bosnia-Herzegovina." *Dialectical Anthropology* 30(3–4): 177–99.

Jasarevic, Larisa. 2014. "Speculative technologies: Debt, love, and divination in a transnationalizing market." *WSQ: Women's Studies Quarterly* 1(2): 261–77.

Jaymalin, Mayen. 2009. "Afghanistan helicopter crash: 10 Pinoys killed." *Philippine Star,* July 21. http://www.philstar.com/headlines/488413/afghanistan-helicopter-crash-10 -pinoys-killed.

———. 2010. "Over 5,000 undocumented Pinoys working in Afghanistan." *Philippine Star,* July 5. http://www.philstar.com/headlines/590129/over-5000-undocumented-pinoys -working-afghanistan.

Jennings, Kathleen M. 2010. "Unintended consequences of intimacy: Political economies of peacekeeping and sex tourism." *International Peacekeeping* 17(2): 229–43.

Jennings, Kathleen M., and Morten Boas. 2015. "Transactions and interactions: Everyday life in the peacekeeping economy." *Journal of Intervention and Statebuilding* 9(3): 281–95.

Jia, Sen, Thomas Lansdall-Welfare, Saatviga Sudhahar, Cynthia Carter, and Nello Cristianini. 2016. "Women are seen more than heard in online newspapers." *PLOS One* 11(2): 1–11.

Joachim, Jutta, and Andrea Schneiker. 2012. "Of 'true professionals' and 'ethical warriors': A gender-discourse analysis of private military and security companies." *Security Dialogue* 43(6): 495–512.

———. 2015. "The license to exploit: PMSCs, masculinities, and third country nationals." In *Gender and private security in global politics*, edited by Maya Eichler, 114–30. New York: Oxford University Press.

Johnson, Chalmers. 2004. *The sorrows of empire: Militarism, secrecy, and the end of the republic*. New York: Metropolitan Books.

Jones, Craig. 2015. "Frames of law: Targeting advice and operational law in the Israeli military." *Environment and Planning D: Society and Space* 33(4): 676–96.

———. 2016. "Lawfare and the juridification of late modern war." *Progress in Human Geography* 40(2): 221–39.

Kammerer, Peter. 2003. "Envoy goes to Gulf to negotiate jobs for Filipinos." *South China Morning Post*, May 5.

Kaplan, Amy. 2003. *The anarchy of empire in the making of U.S. culture*. Cambridge, MA: Harvard University Press.

———. 2005. "Where is Guantanamo?" *American Quarterly* 57(3): 831–58.

Kaplan, Robert. 2003. "Supremacy by stealth." *Atlantic Monthly* (July–August). http://www.theatlantic.com/magazine/archive/2003/07/supremacy-by-stealth/302760.

Kavinnamannil, Sindhu, and Sam McCahon. 2011. "In the name of progress: Illegal human labor trafficking within government contracts." *Fraud Magazine* (May–June). http://www.fraud-magazine.com/article.aspx?id=4294969377.

Keen, David. 2006. *Endless war? Hidden functions of the "war on terror."* London: Pluto Press.

Kelly, Philip, and Tom Lusis. 2006. "Migration and the transnational habitus: Evidence from Canada and the Philippines." *Environment and Planning A* 38: 831–47.

Kendrick, Randall, Richard Hawkins, and Brian Swan. 2012. "U.S. Army Europe: A strategic enabler for the northern distribution network." *Army Magazine* 62(2): 28–31.

Kern, Alice, and Ulrike Muller-Boker. 2015. "'The middle space of migration.' A case study on brokerage and recruitment agencies in Nepal." *Geoforum* 65: 158–69.

Khalili, Laleh. 2012. *Time in the shadows: Confinement in counterinsurgencies*. Stanford, CA: Stanford University Press.

———. 2018. "The infrastructural power of the military: The geoeconomic role of the US Army Corps of Engineers in the Arabian Peninsula." *European Journal of International Relations* 24(4): 911–33.

Khan, Mirwais, and Sebastian Abbot. 2012. "Resumed NATO convoys mean millions for Taliban." *Military.com*, July 31. http://www.military.com/daily-news/2012/07/31/renewed-nato-convoys-means-millions-for-taliban.html?ESRC=eb.nl.

King, Akil, Zackary Moss, and Afi Pittman. 2014. "Overcoming logistics challenges in East Africa." *Army Sustainment* (January–February). http://www.alu.army.mil/alog/PDF/JANFEB2014/117849.pdf.

Kinsey, Christopher. 2009. *Private contractors and the reconstruction of Iraq: Transforming military logistics*. New York: Routledge.

Kinsey, Christopher, and Mark Erbel. 2011. "Contracting out support services in future expeditionary operations: Learning from the Afghan experience." *Journal of Contemporary European Research* 7(4): 539–60.

Knežević, Gordana. 2017. "'Nothing to be cheerful about': Bosnians, traumatized by war, modern life, turn to antidepressants." *Radio Free Europe, Radio Liberty*, September 28. https://www.rferl.org/a/bosnia-antidepressants-mass-anxiety-postwar-anxiety/28762076.html.

Krahmann, Elke. 2013. "The United States, PMSCs and the state monopoly on violence: Leading the way towards norm change." *Security Dialogue* 44(1): 53–71.

Kreig, Andreas, and Jean-Marc Rickli. 2018. "Surrogate warfare: The art of war in the 21st century." *Defence Studies* 18(2): 113–30.

Kronstadt, Alan. 2011. *Pakistan-U.S. relations: A summary*. Congressional Research Service. October 21. https://pdfs.semanticscholar.org/9beb/cbcbdd1ca661894c216e87 cdc3fd5aafc647.pdf.

Kuchins, Andrew, Thomas Sanderson, and David Gordon. 2009. *The Northern Distribution Network and the modern Silk Road: Planning for Afghanistan's future*. Center for Strategic and International Studies. https://csis-prod.s3.amazonaws.com/s3fs-public /legacy_files/files/publication/091217_Kuchins_NorthernDistNet_Web.pdf.

Kuchins, Andrew, and Shalini Sharan. 2015. "Both epicenter and periphery: U.S. interests in Central Asia." In *China, the United States and the future of Central Asia*, edited by David Denoon, 101–29. New York: New York University Press.

Kurtović, Larisa. 2015. "'Who sows hunger, reaps rage': On protest, indignation and redistributive justice in post-Dayton Bosnia-Herzegovina." *Southeast European and Black Sea Studies* 15(4): 639–59.

Lair, Meredith. 2001. *Armed with abundance: Consumerism and soldiering in the Vietnam War*. Chapel Hill: University of North Carolina Press.

Langley, Anthony. 2010. "Africa Command ISR Initiative Operations (Aii Ops) Contract: Industry Day." PowerPoint presentation. June 11. https://www.fbo.gov/utils/view? id=1c457e4bd2c38a77de7f10add7f4e634.

Lasker, Bruno. 1969. *Filipino immigration to continental United States and to Hawaii*. Chicago: University of Chicago Press. Original work published 1931.

LeBaron, Genevieve. 2014. "Subcontracting is not illegal, but is it unethical? Business ethics, forced labor, and economic success." *Brown Journal of World Affairs* 20(2): 237–49.

Lee-Brago, Pia. 2005. "Iraq: Filipino workers protest working conditions under KBR." *CorpWatch*, May 27. http://www.corpwatch.org/article.php?id=12328.

Legarda Jr., Benito. 1955. "Two and a half centuries of the galleon trade." *Philippine Studies* 3(4): 345–72.

Li, Darryl. 2015. "Offshoring the army: Migrant workers and the U.S. military." *UCLA Law Review* 62: 123–74.

Lin, Weiqiang, Johan Lindquist, Biao Xiang, and S. A. Yeoh. 2017. "Migration infrastructures and the production of migrant mobilities." *Mobilities* 12(2): 167–74.

Lindquist, Johan. 2012. "The elementary school teacher, the thug and his grandmother: Informal brokers and transnational migration from Indonesia." *Pacific Affairs* 85(1): 69–98.

——. 2015. "Of figures and types: Brokering knowledge and migration in Indonesia and beyond." *Journal of the Royal Anthropological Institute* 21(S1): 162–77.

——. 2017. "Brokers, channels, infrastructure: Moving migrant labor in the Indonesian-Malaysian oil palm complex." *Mobilities* 12(2): 212–226.

Lindquist, Johan, Biao Xiang, and Brenda Yeoh. 2012. "Opening the black box of migration: Brokers, the organization of transnational mobility and the changing political economy in Asia." *Pacific Affairs* 85(1): 7–19.

Lipman, Jana. 2008. *Guantanamo: A working class history between empire and revolution*. Berkeley, CA: University of California Press.

Lutz, Catherine. 2006. "Empire is in the details." *American Ethnologist* 33(4): 593–611.

——, ed. 2008. *The bases of empire: The global struggle against U.S. military posts*. New York: New York University Press.

Mann, Geoff, Charmaine Chua, Anja Kanngieser, Mazen Labban, and Deborah Cowen. 2017. "Reading Deborah Cowen's *The deadly life of logistics: Mapping violence in the global trade*." *Political Geography* 61: 263–71.

Mann, Michael. 2012. *The sources of social power. Volume 3: Global empires and revolution, 1890-1945*. New York: Cambridge University Press.

Masood, Salman. 2009. "Bridge attack halts NATO supplies to Afghanistan." *New York Times*, February 3. http://www.nytimes.com/2009/02/04/world/asia/04pstan.html.

Mauldin, Bill. 1968. *Up front*. New York: W. W. Norton. Original work published 1945.

Mayberry, Rory. 2007. "Written testimony." House Committee on Oversight and Government Reform, Subcommittee on National Security and Foreign Affairs. July 26. http://www.globalsecurity.org/military/library/congress/2007_hr/070726-mayberry.pdf.

Mayer, Holly A. 1996. *Belonging to the army: Camp followers and community during the American Revolution*. Columbia, SC: University of South Carolina Press.

McCahon, Sam. 2011. "Written testimony." House Committee on Oversight and Government Reform, Subcommittee on Technology Information Policy, Intergovernmental Relations and Procurement Reform. November 2. https://oversight.house.gov/wp-content/uploads/2012/01/11-2-11_McCahon_TechIP_Testimony_FINAL.pdf.

McDonnell, James, and Ronald Novack. 2004. "Logistics challenges in support of Operation Enduring Freedom." *Army Logistician* 36(5): 9–13.

McFate, Sean. 2015. *The modern mercenary: Private armies and what they mean for world order*. Oxford: Oxford University Press.

McGrath, John. 2007. *The other end of the spear: The tooth to tail ratio (T3R) in modern military operations*. Long War Series Occasional Paper 23. Fort Leavenworth, KS: Combat Studies Institute Press.

McKenna, Dave. 2002. "U.S. military logistics management, privatization, and contractors on the battlefield. What does this all mean?" U.S. Army War College Research Paper. April 9. http://www.dtic.mil/cgi-bin/GetTRDoc?AD=ADA404267.

McKenna, Rebecca Tinio. 2016. *American imperial pastoral: The architecture of US colonialism in the Philippines*. Chicago: University of Chicago Press.

McNabb, Duncan. 2009. *Testimony before the U.S. Senate Committee on Armed Services*. March 17. https://www.gpo.gov/fdsys/pkg/CHRG-111shrg52620/pdf/CHRG-111shrg52620.pdf.

McNulty, John. 2009. "U.S. Army logistics in military operations." Headquarters, Department of the Army. April 27. Copy on file with author.

McQue, Katie. 2017. "Exclusive: Inside Diego Garcia, America's highly secretive military base." *New Internationalist*, March 14. https://newint.org/features/web-exclusive/2017/03/14/inside-diego-garcia-americas-highly-secretive-military-base.

Mendelson, Sarah E. 2005. *Barracks and brothels: Peacekeepers and human trafficking in the Balkans*. Center for Strategic and International Studies. February 1. https://www.csis.org/analysis/barracks-and-brothels.

Merle, Renae. 2006. "Census counts 100,000 contractors in Iraq." *Washington Post*, December 5. http://www.washingtonpost.com/wp-dyn/content/article/2006/12/04/AR2006120401311.html.

Michels, Patrick. 2008. "Private trauma." *Texas Observer*, March 21. https://www.texasobserver.org/2720-private-trauma-preston-wheeler-went-to-iraq-to-make-a-fortune-but-came-home-a-wounded-man.

Miller, Christine. 2012. "Kellogg Brown & Root Services, Inc. v. U.S. (No. 09-351C)." *Leagle*. May 2. http://www.leagle.com/decision/In%20FDCO%2020120503902/KELLOGG%20BROWN%20&%20ROOT%20SERVICES,%20INC.%20v.%20U.S.

Miller, T. Christian. 2006. *Blood money: Wasted billions, lost lives, and corporate greed in Iraq*. New York: Little, Brown.

Miller, Richard. 2004. *The messman chronicles: African-Americans in the U.S. Navy, 1932–1943*. Annapolis, MD: Naval Institute Press.

Minca, Claudio, and Chin-Ee Ong. 2016. "The power of space: The biopolitics of custody and care at the Lloyd Hotel, Amsterdam." *Political Geography* 52: 34–46.

Mitchell, Katharyne. 2010. "Ungoverned space: Global security and the geopolitics of broken windows." *Political Geography* 29(5): 289–97.

Mittelstadt, Jennifer. 2015. *The rise of the military welfare state*. Cambridge, MA: Harvard University Press.

Molland, Sverre. 2012. "Safe migration, dilettante brokers and the appropriation of legality: Lao-Thai 'trafficking' in the context of regulating labor migration." *Pacific Affairs* 85(1): 117–36.

Moore, Adam. 2013. *Peacebuilding in practice: Local experience in two Bosnian towns*. Ithaca, NY: Cornell University Press.

———. 2016a. "The evolving U.S. military presence in Niger." *Conflict Geographies* (blog), August 16. https://conflictgeographies.com/2016/08/16/the-evolving-u-s-military-presence-in-niger.

———. 2016b. "Logistics contracting and U.S. military operations in Niger and Cameroon." *Conflict Geographies* (blog), September 29. https://conflictgeographies.com/2016/09/29/logistics-contracting-and-u-s-military-operations-in-niger-and-cameroon.

———. 2017. "U.S. military logistics outsourcing and the everywhere of war." *Territory, Politics, Governance* 5(1): 5–27.

———. 2018. "Military contracting and the labor of force projection." In *Geographies of Power*, edited by John Agnew and Mat Coleman, 332–46. New York: Edward Elgar.

———. Forthcoming. "Localizing peacebuilding: The Arizona market and the evolution of U.S. military peacekeeping priorities in Bosnia." *Journal of Intervention and Statebuilding*.

Moore, Adam, and James Walker. 2016. "Tracing the U.S. military's presence in Africa." *Geopolitics* 21(3): 686–716.

Morrison, Kenneth. 2016. *Sarajevo's Holiday Inn on the frontline of politics and war*. London: Palgrave Macmillan.

Mount, Mike. 2008. "Suit: KBR forced Nepali men to work against will in Iraq." *CNN.com*, August 28. http://www.cnn.com/2008/WORLD/asiapcf/08/28/kbr.nepal.workers.

Mujkić, Asim. 2016. "Bosnian days of reckoning: Review of the sequence of protests in Bosnia and Herzegovina 2013–14, and future prospects." *Southeastern Europe* 40(2): 217–42.

Musheno, Michael, and Susan R. Ross. 2008. *Deployed: How reservists bear the burden of Iraq*. Ann Arbor, MI: University of Michigan Press.

Mynes, Leona. 2010. "Filipinos celebrate independence, share culture with Guantanamo." *Navy News Service*, June 18. http://www.navy.mil/submit/display.asp?story_id=54159.

Nagl, John, and Jonathan DeMella. 2011. "A primer on prime contractor-sub contractor disputes under federal contracts." *Procurement Lawyer* 46(2): 12–16.

"News briefs." 1996. *The Talon*, April 19. http://www.dtic.mil/bosnia/talon/tal19960419.pdf.

Nickel, Shawn. 2014. "Romania airbase replaces transit center Manas." U.S. Air Force. August 22. http://www.af.mil/News/ArticleDisplay/tabid/223/Article/494562/romania-air-base-replaces-transit-center-manas.aspx.

Nolan, Richard. 2010. *Memorandum for all contractors in Iraq: Contractor compliance with US/TCN/GOI laws and contract demobilization*. CENTCOM Contracting Command. July 20. Copy on file with author.

Obama, Barak. 2015. *National Security Strategy*. Washington, DC: White House. http://nssarchive.us/wp-content/uploads/2015/02/2015.pdf.

——. 2016. "Remarks by the president on the Ebola outbreak." https://obamawhitehouse.archives.gov/the-press-office/2014/09/16/remarks-president-ebola-outbreak.

O'Connell, Patricia. 2003. "A Philippine foothold in Iraq." *Bloomberg*, May 27. https://www.bloomberg.com/news/articles/2003-05-27/a-philippine-foothold-in-iraq.

Oldenziel, Ruth. 2011. "Islands: The United States as a networked empire." In *Entangled geographies: Empire and technopolitics in the global Cold War*, edited by Gabrielle Hecht, 13–42. Cambridge, MA: MIT Press.

Orbeta Jr., Aniceto, and Michael Abrigo. 2009. *Philippine international labor migration in the past 30 years: Trends and prospects*. Discussion Paper Series No. 2009-33. Philippine Institute for Development Studies. https://dirp3.pids.gov.ph/ris/dps/pidsdps0933.pdf.

Owens, John. 2007. "Written testimony." House Committee on Oversight and Government Reform, Subcommittee on National Security and Foreign Affairs. July 26. http://www.globalsecurity.org/military/library/congress/2007_hr/070726-owens.pdf.

Paddock, Richard. 2006. "The overseas class." *Los Angeles Times*. April 20. http://articles.latimes.com/2006/apr/20/world/fg-remit20.

Paglen, Trevor. 2009. *Blank spots on the map: The dark geography of the Pentagon's secret world*. New York: Penguin Group.

Paley, Amit. 2008. "In Iraq, 'a prison full of innocent men.'" *Washington Post*, December 6. http://www.washingtonpost.com/wp-dyn/content/article/2008/12/05/AR2008120503906.html.

Palmer, Herman T. 1999. "More tooth, less tail: Contractors in Bosnia." *Army Logistician* 31(5): 6–9.

Pangilinan, Francis. 2004. "Why Manila left Iraq early." *Far Eastern Economic Review*, July 29: 24.

Pargan, Mehmed. 2009. "Lukavački odgovor na recesiju" [Lukavac residents' answer to the recession]. *Slobodna Bosna*, April 2.

——. 2010. "U Američkim bazama u Iraku i Afganistanu radi 5.000 Tuzlanskih mladića" [5,000 young people from Tuzla work on American bases in Iraq and Afghanistan]. *Slobodna Bosna*, March 18.

Pattison, James. 2014. *The morality of private war: The challenge of private military and security companies*. Oxford: Oxford University Press.

Penney, Joe, Eric Schmitt, Rukmini Callimachi, and Christoph Koettl. 2018. "C.I.A. drone mission, curtailed by Obama, is expanded in Africa under Trump." *New York Times*, September 9. https://www.nytimes.com/2018/09/09/world/africa/cia-drones-africa-military.html?action=click&module=Top%20Stories&pgtype=Homepage.

Perlez, Jane, and Helene Cooper. 2010. "Signaling tensions, Pakistan shuts NATO route." *New York Times*, September 30. http://www.nytimes.com/2010/10/01/world/asia/01peshawar.html.

Peters, Heidi M., Moshe Schwartz, and Lawrence Kapp. 2015. *Department of Defense contractor and troop levels in Iraq and Afghanistan: 2007–2015*. Congressional Research Service. December 1. https://digital.library.unt.edu/ark:/67531/metadc795868/m1/1/high_res_d/R44116_2015Dec01.pdf.

——. 2016. *Department of Defense contractor and troop levels in Iraq and Afghanistan: 2007–2016*. Congressional Research Service. August 15. https://news.usni.org/wp-content/uploads/2016/08/R44116.pdf.

Philippine Overseas Employment Administration. n.d. "Bilateral labor agreements (land-based)." http://www.poea.gov.ph/laborinfo/bLB.html.

——. 2011. *Governing Board Resolution No. 5*. September 2. http://www.poea.gov.ph/gbr /2011/5.pdf.

Philippine Statistics Authority. 2016a. "Total number of OFWs estimated at 2.4 million (results from the 2015 Survey on Overseas Filipinos)." April 14. https://psa.gov.ph /content/total-number-ofws-estimated-24-million-results-2015-survey-overseas -filipinos.

——. 2016b. "2015 Survey on Overseas Filipinos." May 18. https://psa.gov.ph/content /2015-survey-overseas-filipinos-0.

Phinney, David. 2005. "Blood, sweat and tears: Asia's poor build U.S. bases in Iraq." *CorpWatch*, October 3. http://www.corpwatch.org/article.php?id=12675.

——. 2006. "A U.S. fortress rises in Baghdad: Asian workers trafficked to build world's largest embassy." *CorpWatch*, October 17. http://www.corpwatch.org/article.php?id =14173.

Pietrucha, Mike. 2012. "Logistical fratricide." *Armed Forces Journal*, February 1. http:// armedforcesjournal.com/logistical-fratricide.

Pillsbury, James. 2010. "Not your father's AMC." *Defense AT&L* (November–December): 2–8. http://www.dau.mil/pubscats/ATL%20Docs/Nov-Dec10/Pillsbury.pdf.

Ploch, Laura. 2011. *Africa Command: U.S. strategic interests and the role of the U.S. military in Africa*. Congressional Research Service. July 22. https://www.fas.org/sgp/crs /natsec/RL34003.pdf.

Poblete, JoAnna. 2014. *Islanders in the empire: Filipino and Puerto Rican laborers in Hawai'i*. Urbana: University of Illinois Press.

Potts, Shaina. 2017. "Displaced sovereignty: U.S. law and the transformation of international financial space." PhD diss., University of California, Berkeley.

Project on Government Oversight. n.d. "Gulf Catering Company for General Trade and Contracting v. KBR (DFAC subcontract payment dispute)." *Federal Contractor Misconduct Database*. https://www.contractormisconduct.org/misconduct/2230/gulf -catering-company-for-general-trade-and-contracting-v-kbr-dfac-subcontract -payment-dispute.

Purviance, S. A. 1907. "Durbar Week at Agra." *Journal of the U.S. Calvary Association* 17(July 1906–July 1907): 624–29.

Quinsaat, Jesse. 1976. "An exercise on how to join the Navy and still not see the world . . ." In *Letters in exile: An introductory reader on the history of Pilipinos in America*, edited by Jesse Quinsaat, 96–110. Los Angeles: UCLA Asian American Studies Center.

Rackuaskas, Ed. 2008. "Subsistence." Briefing. Defense Supply Center Philadelphia. February 25. http://www.quartermaster.army.mil/jccoe/operations_directorate/reserve _component/RC_Workshop2008/presentations/Guest_Speaker/DSCP%20 BRIEF%20.pdf.

Rasor, Dina, and Robert Bauman. 2007. *Betraying our troops: The destructive results of privatizing war*. New York: Palgrave Macmillan.

Raugh, Harold E. 2010. *Operation Joint Endeavor, V Corps in Bosnia-Herzegovina, 1995– 1996: An oral history*. Fort Leavenworth, KS: Combat Studies Institute Press.

Raustiala, Kal. 2009. *Does the constitution follow the flag? The evolution of territoriality in American law*. New York: Oxford University Press.

Raz, Guy. 2007. "U.S. builds air base in Iraq for the long haul." *NPR*, October 12. http:// www.npr.org/templates/story/story.php?storyId=15184773.

Reeve, Richard, and Zoe Pelter. 2014. *From new frontier to new normal: Counter-terrorism operations in the Sahel-Sahara*. Remote Control Project, Oxford Research Group. https://www.oxfordresearchgroup.org.uk/from-new-frontier-to-new-normal -counter-terrorism-operations-in-the-sahel-sahara.

"Regruteri kompanije Fluor stigli u Tuzlu" [Recruiters for Fluor have arrived in Tuzlu]. 2017. *Sodalive.ba*, February 6. http://www.sodalive.ba/drustvo/regruteri-kompanije-fluor-stigli-u-tuzlu.

Reibestein, Jeff. 2015. "Logistics in support of Operation United Assistance: Teamwork, transitions and lessons learned." United States Africa Command. June 19. http://www.africom.mil/NewsByCategory/Article/25458/logistics-in-support-of-operation-united-assistance-teamwork-transition-and-lessons-learned.

Reif, Jasmine. 2014. "SP-MAGTF CR redeploys to Moron, Spain." United States Africa Command. March 5. http://www.hoa.africom.mil/story/7976/sp-magtf-cr-redeploys-to-mor-n-spain.

Reinhart, Joseph R. 2004. *Two Germans in the civil war: The diary of John Daeuble and the letters of Gottfreid Rentschler, 6th Kentucky Volunteer Infantry*. Knoxville, TN: University of Tennessee.

Roberts, Sue. 2014. "Development capital: USAID and the rise of development contractors." *Annals of the Association of American Geographers* 104(5): 1030–51.

Roderick, Ian. 2010. "Considering the fetish value of EOD robots: How robots save lives and sell war." *International Journal of Cultural Studies* 13(3): 235–53.

Rodriguez, David. 2015. *Statement before the Senate Armed Services Committee*. March 26. http://www.armed-services.senate.gov/imo/media/doc/Rodriguez_03-26-15.pdf.

———. 2016. *Testimony before the Senate Armed Services Committee*. March 8. http://www.africom.mil/media-room/document/28035/2016-posture-statement.

Rodriguez, Robyn Magalit. 2010. *Migrants for export: How the Philippine state brokers labor to the world*. Minneapolis: University of Minnesota Press.

Rohde, David. 2004. "The struggle for Iraq: Foreign labor; Indians who worked in Iraq complain of exploitation." *New York Times*. May 7. http://www.nytimes.com/2004/05/07/world/struggle-for-iraq-foreign-labor-indians-who-worked-iraq-complain-exploitation.html.

Roston, Aram. 2009. "How the US funds the Taliban." *The Nation*, November 11. https://www.thenation.com/article/how-us-funds-taliban.

Ruiz, Ramona. 2012. "Job scam strands workers in Kandahar." *The National*, May 17. https://www.thenational.ae/uae/jobs-scam-strands-workers-in-kandahar-1.362648.

Runstrom, Mark. 2010. "Joint Staff OCS update." PowerPoint presentation. Joint Staff, Department of Defense. July. Copy on file with author.

Ryan, Maria. 2011. "'War in countries we are not at war with': The 'war on terror' on the periphery from Bush to Obama." *International Politics* 48(2–3): 364–389.

Sadler, Anne G., Brenda M. Booth, James C. Torner, and Michelle A. Mengeling. 2017. "Sexual assault in the US military: A comparison of risk in deployed and non-deployed locations among Operation Enduring Freedom/Operation Iraqi Freedom active component and Reserve/National Guard servicewomen." *American Journal of Industrial Medicine* 60(11): 947–55.

Santora, Marc. 2009. "Big bases are a part of Iraq, but a world apart." *New York Times*, September 8. http://www.nytimes.com/2009/09/09/world/middleeast/09bases.html?_r=0.

Sauer, Frank, and Niklas Schornig. 2012. "Killer drones: The 'silver bullet' of democratic warfare?" *Security Dialogue* 43(4): 363–80.

Scahill, Jeremy. 2008. *Blackwater: The rise of the world's most powerful mercenary army*. New York: Nation Books.

———. 2015. "Germany is the tell-tale heart of America's drone war." *The Intercept*, April 17. https://theintercept.com/2015/04/17/ramstein.

Schmitt, Eric, and Mark Mazzetti. 2008. "Secret order lets U.S. raid Al Qaeda." *New York Times*, November 9. http://www.nytimes.com/2008/11/10/washington/10military .html?pagewanted=all&_r=0.

Schooner, Steven L., and Collin Swan. 2012. "Dead contractors: The un-examined effect of surrogates on the public's casualty sensitivity." *Journal of National Security Law and Policy* 6(1): 11–58.

Schwartz, Moshe. 2010. *Department of Defense contractors in Iraq and Afghanistan: Background and analysis*. Congressional Research Service. July 2. https://fas.org/sgp/crs /natsec/R40764.pdf.

Scully, Eileen. 2001. *Bargaining with the state from afar: American citizenship in treaty port China, 1844–1942*. New York: Columbia University Press.

Seck, Hope. 2015. "Crisis response Marines test 3 Africa staging bases." *Marine Corps Times*, July 10. http://www.marinecorpstimes.com/story/military/2015/07/09/marine -corps-staging-bases-africa/29879357.

Secretary of State. 2007. "Establishing a northern ground line of communication." Department of State cable (Confidential). November 6. https://wikileaks.org/plusd/cables /07STATE153352_a.html.

Semple, Kirk. 2005. "G.I.'s deployed in Iraq desert with lots of American stuff." *New York Times*, August 13. http://www.nytimes.com/2005/08/13/world/middleeast/gis -deployed-in-iraq-desert-with-lots-of-american-stuff.html?_r=0.

Serafino, Nina M. 2001. *Colombia: Plan Colombia legislation and assistance (FY2000– FY2001)*. Congressional Research Service. July 5. https://fas.org/asmp/resources /govern/crs-RL30541.pdf.

Serquina Jr., Oscar. 2016. "'The greatest workers of the world': Philippine labor outmigration and the politics of labeling in Gloria Macapagal-Arroyo's presidential rhetoric." *Philippine Political Science Journal* 37(3): 207–27.

Sharma, Miriam. 1984. "Labor migration and class formation among Filipinos in Hawaii, 1906–1946." In *Labor immigration under capitalism: Asian workers in the United States before World War II*, edited by Lucie Cheng, and Enda Bonacich, 579–616. Berkeley: University of California Press.

Shaw, Ian. 2013. "Predator empire: The geopolitics of U.S. drone warfare." *Geopolitics* 18(3): 536–59.

———. 2016. *Predator empire: Drone warfare and full spectrum dominance*. Minneapolis: University of Minnesota Press.

Shaw, Ian, and Majed Akhter. 2012. "The unbearable humanness of drone warfare in FATA, Pakistan." *Antipode* 44(4): 1490–509.

Shaw, Martin. 2002. "Risk transfer militarism, small massacres, and the historic legitimacy of war." *International Relations* 16(3): 343–60.

———. 2005. *The new Western way of war*. Cambridge: Polity.

Sheehan, Neil. 1967. "Officials in U.S. irked by report of low ratio of combat troops." *New York Times*, July 13.

Shor, Eran, Arnout van de Rijtb, Alex Miltsova, Vivek Kulkarnib, and Steven Skienab. 2015. "A paper ceiling: Explaining the persistent underrepresentation of women in printed news." *American Sociological Review* 80(5): 960–84.

Shorrock, Tim. 2008. *Spies for hire: The secretive world of intelligence outsourcing*. New York: Simon and Schuster.

Simbulan, Roland. 1983. *The bases of our insecurity: A study of the U.S. military bases in the Philippines*. Manila: BALAI.

Simon, Bob. 2014. "Pink Panthers: Daring heists rake in half-a-billion dollars." *60 Minutes*. March 23. http://www.cbsnews.com/news/pink-panthers-gang-of-diamond-thieves.

Simpson, Cam. 2005a. "Desperate for work, lured into danger." *Chicago Tribune*, October 9.

———. 2005b. "Into a war zone, on a deadly road." *Chicago Tribune*. October 10.

———. 2018. *The girl from Kathmandu: Twelve dead men and a woman's quest for justice.* New York: Harper Collins.

Singer, Peter W. 2008. *Corporate warriors: The rise of the privatized military industry.* Ithaca, NY: Cornell University Press.

———. 2009. *Wired for war: The robotics revolution and conflict in the twenty-first century.* London: Penguin.

Skokić, Edin. 2008. "Na današnji dan život je izgubio Nedim Nuhanović" [Nedim Nuhanović lost his life on this day]. *Sodalive.ba*, June 11. http://www.sodalive.ba/aktuelnosti /lukavac-grad/na-danasnji-dan-zivot-je-izgubio-nedim-nuhanovic.

Slavnić, Aleksandra. 2013. "Mladi u BiH—Avganistan ili luksuzni kruzer?" [Youth in BiH– Afghanistan or luxury cruise ship]. *Sodalive.ba*, March 11. http://www.sodalive.ba /drustvo/mladi-u-bih-avganistan-ili-luksuzni-kruzer.

Sloop, Rick. 2016. *Fluor support to Operation United Assistance.* Presentation to the Society of American Military Engineers—Tampa Bay Post. February 10. Copy of slide deck on file with author.

Smirl, Lisa. 2015. *Spaces of aid: How cars, compounds and hotels shape humanitarianism.* London: Zed Books.

———. 2016. "'Not welcome at the Holiday Inn': How a Sarajevo hotel influenced geopolitics." *Journal of Intervention and Statebuilding* 10(1): 32–55.

Smith, Charles M. 2012. *War for profit: Army contracting vs. supporting the troops.* New York: Algora.

Smith, Neil. 2003. *American empire: Roosevelt's geographer and the prelude to globalization.* Berkeley: University of California Press.

Snell, Angela. 2011. "The absence of justice: Private military contractors, sexual assault, and the U.S. government's policy of indifference." *University of Illinois Law Review* (3): 1125–64.

Sopko, John. 2014. *Quarterly report to the United States Congress.* Special Inspector General for Afghanistan Reconstruction. October 30. https://www.sigar.mil/pdf /quarterlyreports/2014-10-30qr.pdf.

Soriano, Cesar G. 1996. "SJA explains black marketing." *The Talon*, May 24. http://www .dtic.mil/bosnia/talon/tal19960524.pdf.

Stachowitsch, Saskia. 2013. "Military privatization and the remasculinization of the state: Making the link between outsourcing of military security and gendered state transformations." *International Relations* 27(1): 74–94.

———. 2015. "The reconstruction of masculinities in global politics: Gendering strategies in the field of private security." *Men and Masculinities* 18(3): 363–86.

Stafford, Darlene E., and James M. Jondrow. 1996. *A survey of privatization and outsourcing initiatives.* Center for Naval Analyses. http://www.dtic.mil/cgi-bin/GetTRDoc ?AD=ADA362409.

Stanger, Allison. 2009. *One nation under contract: The outsourcing of American power and the future of foreign policy.* New Haven, CT: Yale University Press.

Stanton, Shelby. 2003. *The rise and fall of an American army: U.S. ground forces in Vietnam, 1963–1973.* New York: Presidio Press.

Stillman, Sarah. 2011. "The invisible army." *New Yorker*, June 6. http://www.newyorker .com/magazine/2011/06/06/the-invisible-army.

Stoler, Ann Laura. 2001. "Tense and tender ties: The politics of comparison in North American history and (post)colonial studies." *Journal of American History* 88(3): 829–65.

——. 2002. *Carnal knowledge and imperial power: Race and the intimate in colonial rule.* Berkeley: University of California Press.

——. 2016. *Duress: Imperial durabilities in our times.* Durham, NC: Duke University Press.

Stollenwerk, Michael F. 1998. *LOGCAP: Can battlefield privatization and outsourcing create tactical synergy?* Fort Leavenworth, KS: School of Advanced Military Studies.

Streitfeld, David. 2014. "Engineers allege hiring collusion in Silicon Valley." *New York Times*, February 28. https://www.nytimes.com/2014/03/01/technology/engineers-allege-hiring-collusion-in-silicon-valley.html?mcubz=3&_r=0.

Sullivan, Julie. 2010. "Ms. Sparky aims at KBR, electrifies war-contractor scrutiny with blog." *The Oregonian*, July 11. http://www.oregonlive.com/news/index.ssf/2010/07/ms_sparky_aims_at_war_contract.html.

Task Force Currahee. 2014. *Afghan commander AAR book.* https://docplayer.net/43190754-Afghan-commander-aar-book-currahee-edition-april-2014.html.

Tellis, Ashley. 2011. *Managing frenemies: What should the United States do about Pakistan.* CERI Strategy Papers No. 13. October 24. http://carnegieendowment.org/files/frenemies2011.pdf.

Terry, William. 2014. "The perfect worker: Discursive makings of Filipinos in the workplace hierarchy of the globalized cruise industry." *Social and Cultural Geography* 15(1): 73–93.

Thomas, Kevin. 2017. *Contract workers, risk, and the war in Iraq: Sierra Leonean labor migrants at U.S. military bases.* Montreal: McGill-Queen's Press.

Thorpe, George. 1997. *Pure logistics.* Newport, RI: Naval War College Press. Original work published 1917.

Thrasher, John. 1993. Subcontractor dispute remedies: Asserting subcontractor claims against the federal government." *Public Contract Law Journal* 23(1): 39–104.

Tierney, John. 2010. *Warlord, Inc.: Extortion and corruption along the U.S. supply chain in Afghanistan.* Subcommittee on National Security and Foreign Affairs, Committee on Oversight and Government Reform, U.S. House of Representatives. http://www.cbsnews.com/htdocs/pdf/HNT_Report.pdf.

Tillson, John C. 1997. *The role of external support in total force planning.* Alexandria, VA: Institute for Defense Analyses.

Tokach, Wendy R. 1997. "Be not afraid." *The Talon*, May 9. http://www.dtic.mil/bosnia/talon/tal19970509.pdf.

Traas, Adrian G. 2010. *Engineers at war.* Washington, DC: Center of Military History.

Tregaskis, Richard. 1975. *Southeast Asia: Building the bases.* Washington, DC: Government Printing Office.

Trevithick, Joseph. 2017. "The U.S. military's incident in Niger has been a long time coming." *The Drive*, October 20. http://www.thedrive.com/the-war-zone/15326/the-u-s-militarys-incident-in-niger-has-been-a-long-time-coming.

Trilling, David. 2011. "Northern distribution nightmare." *Foreign Policy*, December 6. http://foreignpolicy.com/2011/12/06/northern-distribution-nightmare.

Turse, Nick. 2017. "The war you've never heard of." *Vice News*, May 18. https://news.vice.com/story/the-u-s-is-waging-a-massive-shadow-war-in-africa-exclusive-documents-reveal.

——. 2018. "U.S. military says it has a 'light footprint' in Africa. These documents show a vast network of bases." *The Intercept.* December 1. https://theintercept.com/2018/12/01/u-s-military-says-it-has-a-light-footprint-in-africa-these-documents-show-a-vast-network-of-bases/.

Tyler, Carissa N. 2012. "Limitations of the contingency contracting framework: Finding effective ways to police foreign subcontractors in Iraq and Afghanistan." *Public Contract Law Journal* 41(2): 453–72.

Tyner, James. 2003. "Globalization and the geography of labor recruitment firms in the Philippines." *Geography Research Forum* 23: 78–95.

———. 2005. *Iraq, terror, and the Philippines' will to war*. Lanham, MD: Rowman and Littlefield.

Tyson, Ann S. 2006. "Possible Iraq deployments would stretch reserve force." *Washington Post*, November 5. http://www.washingtonpost.com/wp-dyn/content/article/2006/11/04/AR2006110401160.html.

Tyson, James. 1986. "Labor talks fail to resolve strike at U.S. bases." *Associated Press*, March 27.

United Nations. 2014. *Security Council Resolution 2177*. http://www.ifrc.org/docs/IDRL/UN%20SC%20Res.pdf.

"US hires 400 Filipinos to build prison for Al Qaeda fighters: Reports." 2002. *Agence France Presse*, March 28.

U.S. Africa Command. 2010. *Performance work statement for Africa Command ISR Initiative Operations (Aii Ops)*. https://www.fbo.gov/utils/view?id=384c28d8edd1fb56856f206d9008d8f8.

U.S. Air Forces in Europe and Air Forces Africa A4A7K. 2015. *Justification and approval for other than full and open competition*. https://www.fbo.gov/utils/view?id=844b7ec0eb38a934612c164d7482d2ea.

U.S. Army. 2015. *Logistics Civil Augmentation Program, Industry Day, LOGCAP V*. October 29. https://www.fbo.gov/utils/view?id=8eaf211d8236a5774f821c7df7fd2226.

U.S. Army Audit Agency. 2013. *Cost-sharing: Logistics support, services, and supplies. U.S. forces-Afghanistan (Audit Report: A-2013-0110-MTE)*. https://slideblast.com/cost-sharing-logistics-support-services-and-supplies_5955d3c91723ddcb2b410fc2.html.

U.S. Court of Appeals, Fifth Circuit. 1992. *Romarico S. More et al v. Intelcom Support Services Inc (No. 91-1325)*. http://www.ca5.uscourts.gov/opinions%5Cpub%5C91/91-1325.0.wpd.pdf.

U.S. Transportation Command. 2012. *2012 Annual report*. Copy on file with author.

———. 2016. *Directorate of Acquisition*. October 31. Copy on file with author.

Vandiver, John. 2013. "Troops fill in for striking Djibouti workers." *Stars and Stripes*, July 16. https://www.stripes.com/news/africa/troops-help-fill-in-for-striking-djibouti-workers-1.230579#.WaiCEsh942w.

———. 2014. "U.S. crisis response Marines mobilized for North Africa threat." *Stars and Stripes*, May 14. http://www.stripes.com/news/us-crisis-response-marines-mobilized-for-north-africa-threat-1.282966.

Ver, I. Wildredo. 2008. "The Fil-Am general's mission." *ABS-CBN News*, May 9. http://news.abs-cbn.com/pinoy-migration/05/09/08/fil-am-general%E2%80%99s-mission.

Verkuil, Paul R. 2007. *Outsourcing sovereignty: Why privatization of government functions threatens democracy and what we can do about it*. New York: Cambridge University Press.

"Vietnam: How business fights the war by contract." 1965. *Business Week*, March 5. Quoted in Steven J. Zamparelli. 1999. "Contractors on the battlefield: What have we signed up for?" Master's thesis, Air War College, Air University. http://www.dtic.mil/cgi-bin/GetTRDoc?Location=U2&doc=GetTRDoc.pdf&AD=ADA395505.

Vine, David. 2009. *Island of shame: The secret history of the U.S. military base on Diego Garcia*. Princeton, NJ: Princeton University Press.

———. 2015. *Base nation: How U.S. military bases abroad harm America and the world*. New York: Metropolitan Books.

Voelz, Glenn J. 2009. "Contractors and intelligence: The private sector in the intelligence community." *International Journal of Intelligence and Counterintelligence* 22(4): 586–613.

Waddell, Steve R. 2009. *United States Army logistics: From the American Revolution to 9/11*. Santa Barbara, CA: Praeger Security International.

Walker, Demetria. 2009. "Securing Khabari crossing in Kuwait." *Army Logistician* 41(1): 2. http://www.alu.army.mil/alog/issues/JanFeb09/secure_khabari.html.

Waite, Louise. 2009. "A place and space for a critical geography of precarity." *Geography Compass* 3(1): 412–33.

Waits, Elizabeth. 2006. "Avoiding the 'legal Bermuda triangle': The Military Extraterritorial Jurisdiction Act's unprecedented expansion of U.S. criminal jurisdiction over foreign nationals." *Arizona Journal of International and Comparative Law* 23(2): 493–542.

Warren, Brittany. 2012. "'If you have a zero-tolerance policy, why aren't you doing anything?': Using the Uniform Code of Military Justice to combat trafficking abroad." *George Washington Law Review* 80(4): 1269–70.

Weizman, Eyal. 2012. *The least of all possible evils: Humanitarian violence from Arendt to Gaza*. London: Verso.

Whelan, John, and George Gnoss. 1968. "Government contracts: Subcontractors and privity." *William and Mary Law Review* 10(1): 80–117.

Whitney, Lance. 2015. "Apple, Google, others settle antipoaching lawsuit for $415 million." *C|Net*, September 3. https://www.cnet.com/news/apple-google-others-settle-anti-poaching-lawsuit-for-415-million.

Whitlock, Craig. 2011. "U.S. turns to other routes to supply Afghan war as relations with Pakistan fray." *Washington Post*, July 2. https://www.washingtonpost.com/world/national-security/us-turns-to-other-routes-to-supply-afghan-war-as-relations-with-pakistan-fray/2011/06/30/AGfflYvH_story.html?utm_term=.82b4010a509d.

———. 2012a. "Leader of Mali military coup trained in US." *Washington Post*, March 24. http://www.washingtonpost.com/world/nationalsecurity/ leader-of-mali-military-coup-trained-in-us/2012/03/23/gIQAS7Q6WS_story.html.

———. 2012b. "U.S. trains African soldiers for Somalia mission." *Washington Post*, May 13. https://www.washingtonpost.com/world/national-security/us-trains-african-soldiers-for-somalia-mission/2012/05/13/gIQAJhsPNU_story.html.

Williams, Allison. 2011. "Enabling persistent presence? Performing the embodied geopolitics of the Unmanned Aerial Vehicle assemblage." *Political Geography* 30(7): 381–90.

Williams, Brian. 2010. "The CIA's covert Predator drone war in Pakistan, 2004–2010: The history of an assassination campaign." *Studies in Conflict and Terrorism* 33(10): 871–92.

Williams, Nick. 1987. "Sailors' dollars, Filipinos' wages: For city outside Subic Naval Base, there was only one voting issue." *Los Angeles Times*, February 3. http://articles.latimes.com/1987-02-03/news/mn-766_1_military-bases.

Wilson, Mark R. 2006. *The business of Civil War: Military mobilization and the state, 1861–1865*. Baltimore: Johns Hopkins University Press.

Wise, Amanda. 2013. "Pyramid subcontracting and moral detachment: Down-sourcing risk and responsibility in the management of transnational labour in Asia." *Economic and Labor Relations Review* 24(3): 433–55.

Woods, Colleen. 2016. "Building empire's archipelago: The imperial politics of Filipino labor in the Pacific." *Labor: Studies in Working-Class History of the Americas* 13(3–4): 131–52.

World Health Organization. 2014. "Statement on the 1st meeting of the IHR Emergency Committee on the 2014 Ebola outbreak in West Africa." August 8. http://www.who .int/mediacentre/news/statements/2014/ebola-20140808/en.

Xiang, Biao, and Johan Lindquist. 2014. "Migration infrastructure." *International Migration Review* 48(S1): 122–48.

Zamparelli, Steven J. 1999. "Contractors on the battlefield: What have we signed up for?" Master's thesis, Air War College, Air University. http://www.dtic.mil/cgi-bin /GetTRDoc?Location=U2&doc=GetTRDoc.pdf&AD=ADA395505.

Zoroya, Gregg. 2012. "VA finds sexual assault more common in warzones." *USA Today*, December 26. https://www.usatoday.com/story/news/nation/2012/12/26/va-finds -sexual-assaults-more-common-in-war-zones/1793253.

Zulueta, Johanna. 2012. "Living as migrants in a place that was once home: The Nisei, the US bases, and Okinawan society." *Philippine Studies: Historical and Ethnographic Viewpoints* 60(3): 367–90.

Index

CPSIA information can be obtained
at www.ICGtesting.com
Printed in the USA
LVHW110428170920
666195LV00001B/61